THE ESSENTIAL Sous Vide Cookbook

THE ESSENTIAL Sous Vide Cookbook

Modern Meals for the Sophisticated Palate

Sarah James

ROCKRIDGE PRESS

For general information on our other products and services or to obtain technical support, please contact our Customer Care Department within the United States at (866) 744-2665, or outside the United States at (510) 253-0500.

Rockridge Press publishes its books in a variety of electronic and print formats. Some content that appears in print may not be available in electronic books, and vice versa.

Interior photographs: p. 36 © Darren Muir/Stocksy,
All others photographs courtesy of Paul Palop

Illustrations by Tom Bingham © 2016

ISBN: Print 978-1-62315-808-8 | eBook 978-1-62315-809-5

CONTENTS

INTRODUCTION

A FEW YEARS AGO when I was the food editor at the DIY site Instructables.com, the CEO, an avid sous vide lover, introduced me to the cooking process. At the time, the gorgeous temperature controllers and circulators for home cooks that we take for granted today did not exist. Home chefs wanting to cook sous vide had to hack together their own temperature controllers and circulators if they wanted to replicate the restaurant-quality food that precision cooking produces. There were plenty of how-tos and recipes shared on Instructables, but it was a fairly niche and esoteric field, and although I knew a lot about the subject, I wasn't too keen on building my own. After a few circulators hit the market through Kickstarter, I started to pay more attention. Then, my brother bought me my first circulator (a Sansaire I still rely on). Let's just say what had been a mild interest turned into an obsession.

I did lot of research, looking for recipes and tips, and discovered that most of the available information was very technical and pretty hard to digest. Soon, because I was cooking most of my meals sous vide, I created Sousvidely.com, a blog dedicated to making sous vide recipes accessible to anyone wanting to learn more about the technique. My own experience showed me how easy it was to use sous vide to produce amazing meals, so I strove to create an approachable, easy-to-understand resource with great information and recipes to help people like me. I've learned so much since then, about food safety, the effects of time and temperature on foods, what works, and what doesn't. I'm excited to be able to pass that knowledge on to you in this book.

There are two reasons I strongly believe everyone should learn about sous vide: It can create amazing, repeatable results that are irreproducible by traditional cooking methods, and it's so very easy—it'll revolutionize the way you think about the preparation and timing of a meal. One of the great joys of cooking sous vide is how little hands-on time it involves. While traditional cooking methods may take less time from start to finish, sous vide requires far less involvement in the actual cooking process and cleanup, while producing a higher-quality result every time. This makes it the perfect solution for busy cooks, whether you're a parent juggling a million responsibilities or a chef cooking for large parties with multiple dishes to produce. It takes less than five minutes to prepare many of the dishes in this book, and the rest of your time can be spent taking the dog for a walk, playing with your kids, drinking a glass of wine, and preparing the rest of the meal.

Plus, cooking sous vide is really fun. Not only do you get to experiment with turning tougher cuts of meat into amazingly tender meals, it's also very impressive. One of the proudest moments of my early sous vide career was bringing a full brisket that had been cooked for 36 hours to a party for the host to finish on the grill. Everyone there agreed it was the best brisket they'd ever had, and it only took me five minutes of hands-on cooking time!

I'm excited to share my passion for cooking sous vide with you. What you'll discover in this book are simple explanations of the hows and whys of sous vide, easy-to-read time and temperature charts, and fun and inspiring recipes to take your cooking skills to the next level. I know you'll find that sous vide will truly simplify your life in the kitchen, just as it did mine.

PART 1

GETTING TO KNOW

SOUS VIDE

1

BEHIND THE SCENES

In Hot Water

Chefs started experimenting with sous vide, as we know it today, in the 1970s. Michelin-star restaurants began incorporating the technique to turn out perfectly cooked meals with minimal effort. As home chefs became savvier, they too wanted to experiment with sous vide techniques, but the cost of professional-grade equipment was often prohibitive. With the rise of the maker movement, more and more crafty cooks started concocting their own temperature-controlled rigs. Luckily for us, popularity grew, and those same hackers eventually developed devices they made available to the public at drastically lower costs. Now sous vide is available to the masses, as is the once-exclusive wealth of information surrounding it.

The basic principle of sous vide is to cook food in a water bath at its ideal target temperature over long periods of time to achieve perfectly cooked food. Traditional cooking methods use high temperatures to cook foods from the outside in. For meats, this creates different zones of doneness as you approach the core—when the inside is the temperature you want, the outer layers will be overcooked. With vegetables, this means you lose the nutrients and essential flavors that leach out at higher temperatures and into water when boiling or blanching. By cooking everything at its ideal target temperature, you create meats that are perfectly cooked throughout the entire cut, and vegetables that retain all the good stuff and showcase surprising flavors. Eggs, whose whites and yolks actually prefer different temperatures, are completely transformed by the precise control you have with sous vide. If you've ever been intimidated by poaching an egg or afraid of breaking a sauce or scrambling a custard, sous vide is your new best friend.

Once you're comfortable with the basics of sous vide, you'll be ready to experiment with more exotic cuts of meat that might otherwise have felt out of your league. Whether that means trying out elk, bison, or boar or just trying to cook salmon right for the first time, sous vide makes once-difficult cooking goals easy to achieve. You'll also be amazed at how the low temperatures and longer cook times can transform cheaper cuts of meat into tender, succulent dishes. In the quiet of your kitchen, right on your countertop, you can turn an inexpensive pork butt into pulled pork that would rival that of any barbecue joint.

New sous vide circulators are small, efficient, and affordable. Compared to electric stoves or microwaves, sous vide circulators take up barely any space and energy and give off very little heat. If you're worried about water consumption, remember that you can reuse the water bath for multiple cooking sessions without a problem. For the budget conscious, entry-level circulators and vacuum sealers can be found for as little as $99 each. If that's still out of range, an entire basic setup can cost as little as a foam cooler and some zip-top bags.

Not only does sous vide promise perfectly cooked food, but it also frees you from the tyranny of the timer. Many foods can remain in the bath for up to an hour past their cooking time without overcooking, making meal prep less of a race and more of an enjoyable ritual. This makes it ideal for the busy parent cooking for their family or anyone throwing a dinner party where guests will be arriving at different times. When you're ready for it, your meal will be ready for you.

Fun Fact: Heston Blumenthal once cooked an entire pig in a hot tub. According to Blumenthal, "We couldn't find a water bath big enough so we went to a hot tub warehouse," he says. "We took the limiters off so it went up to 62°C and we cooked it at this temperature for a day and a half. Then we spit-roasted it, cranking the heat up so it got the browning. It was the best pig that I ever tasted."

BETTER THAN BOILING-IN-A-BAG

You're most likely familiar with the idea of cooking food in a bag. Many companies have been using this method for decades for everything from cooking foolproof rice to reheating complete meals. Sous vide is different. With sous vide, temperatures never reach as high as boiling. It was pioneer chef and scientist Bruno Goussault who introduced the idea that keeping the temperatures lower while cooking yielded a much better result. Goussault also discredited the long-held fear that cooking in bags at low temperatures would most likely lead to botulism. He proved that proper cooling after long cooking kills bacteria the same as cooking at higher temperatures. He also found that sous vide cooking stabilizes the food so that it can be stored longer before serving. Cooking at higher temperatures not only causes the end result to be overcooked in areas but also causes more of the cells to burst, meaning the meats lose their essential liquids. Why take the juices that have cooked off from a turkey and turn them into gravy to then spoon back onto the bird when you don't have to lose that liquid in the first place? Lower temperatures mean juicier meats and more flavorful vegetables.

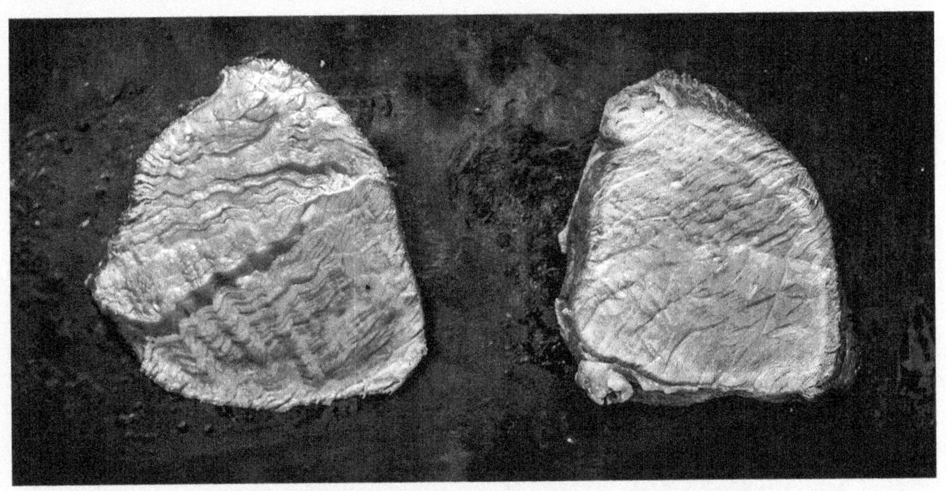

Sous Vide *Traditional*

Not-So-Heavy Machinery

Once upon a time, the microwave was considered futuristic. Now everyone has one. Whereas once it promised to cut cooking times in half and produce delicious and healthy meals, now most are relegated to reheating leftovers. Slow cookers were the next big wave, and while many people still swear by their simplicity, they often produce overcooked, bland dishes and are primarily suited to one-pot meals. Sous vide is truly revolutionary for today's kitchen: It can produce restaurant-quality steaks as well as comfort meals you'd find in most slow cooker cookbooks, but with greater precision and better flavor. Most sous vide devices today can even be controlled by your smartphone, including setting and monitoring temperatures remotely and being notified when your food is done!

Fun Fact: Sous vide was first described by Sir Benjamin Thompson (Count Rumford) in 1799 (although he used air instead of water). It was rediscovered by American and French engineers in the mid-1960s and developed into an industrial food preservation method prior to being used for cooking again.

IMMERSION CIRCULATORS

If you've picked up a sous vide machine recently, chances are it's an immersion circulator from Sansaire, Anova, or Nomiku. Immersion circulators are devices that both heat and circulate the water in sous vide baths. What makes them so convenient is that they're small and affordable, and you can use any container size to suit your needs. Most of these circulators can heat baths up to 8 gallons, with water levels, circulation speeds, and temperature ranges varying between models. Each unit has a different clip design to secure itself to the cooking vessel. The Anova has a rubberized screw that ensures a very firm attachment but requires both hands to connect. The Sansaire uses an oversize clip that's easy to manipulate but feels a little less secure; however, the wide base of the Sansaire unit makes it almost unnecessary. The Nomiku WiFi model has the only front-facing clip, which allows you to read the temperature output from across the room, a feature that is surprisingly handy.

Features of Immersion Circulators on the Market

CIRCULATORS	NOMIKU WIFI	CODLO	JOULE	ANOVA	SANSAIRE
Heating Unit (wattage)	1100W	1000W	1100W	800W	1100W
Temperature Range	32°F to 203°F	68°F to 194°F	Room temp to 209°F	77°F to 210°F	32°F to 212°F
Temperature Accuracy	0.2°F	0.1°F	0.2°F	0.2°F	0.1°F
Weight	2.5 lbs	N/A	1.3 lbs	2.5 lbs	4 lbs
Interface (touchscreen, dial)	Touchscreen + dial	Push button	App only	Touch-screen	Push button + dial
Capacity	5 gallons	N/A	10 gallons	4–5 gallons	6 gallons
WiFi	Y	N	Y	Y	Available spring '17
Bluetooth	N	N	Y	Y	N/A
Integrated App	Y	N	Y	N	Available spring '17
Timer	Yes	Yes	In-App	Yes	N/A
Voltage	120V + 240V	120V + 240V	120V	120V + 220V	110V + 220/240 V
Availability (in-store, online)	Currently online	Currently online	Online	Online	Online + in stores
Circulation Rate	1.8 GPM	Does not circulate	N/A	1.2 GPM	3 GPM
Directional Pump	No	No pump	No	Yes	No
Comes Apart for Cleaning	Yes	N/A	Yes	No	N/A
Method of Attachment to Container	Front-facing clip	Probe	Magnetic foot or clamp	Screw clamp	Clip
Min Water Level	1.5 inches	0.5 inches	1.5 inches	2.5 inches	2.75 inches
Max Water Level	5.5 inches	N/A	8 inches	7.25 inches	6.5 inches
Price	$199	$179	$199	$199	$199
Warranty	1 year	1 year	1 year	1 year	1 year

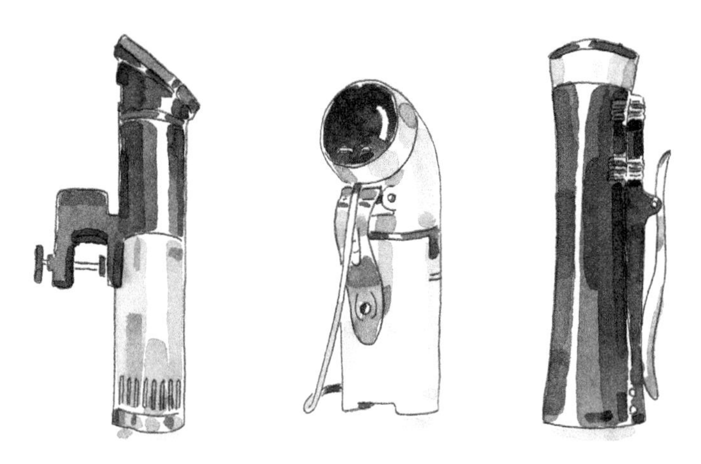

The Nomiku and Anova models both offer Wi-Fi features with apps if you need to be away from the unit when turning it on or off. Personally, I've never found a use for the feature, since I want to make sure my water bath is up to temperature before I put the food in. But the apps come with additional recipes to try, and it does feel pretty neat to be able to control the unit from my phone. All three immersion units are currently priced below $200. All three companies have outstanding customer service if you're ever in need of help. So which is the best? Each of these units currently resides in my kitchen, and while they all get the job done, the Sansaire is the one I reach for the most because of its wide base and easy-to-use clip. New models are hitting the market all the time, so keep yourself informed to choose the best model to suit your needs.

WATER OVENS

If you're serious about sous vide and ready to incorporate it into your everyday routine, you might be interested in investing in a water oven. SousVide Supreme is the current leader in the market, offering three different models of water ovens in 9- or 11-liter sizes, along with a vacuum sealer, proprietary pouches, and a range of accessories. This company has been in the home sous vide game the longest.

Features of Water Ovens on the Market

WATER OVENS	SOUSVANT	MELLOW	SOUSVIDE SUPREME
Heating Unit (wattage)	1000W	1000W	850W
Temperature Range	95°F to 194°F	35°F to 192°F	86°F to 210°F
Temperature Accuracy	0.1°F	1°F	1°F
Interface (touchscreen, dial)	Push button	App only	Push button
Capacity	3 gallons	1.2 gallons	2.6 gallons
Circulation	Yes	Yes	No
Wi-Fi	Yes	Yes	No
Integrated App	No	Yes	No
Timer	No	Yes	Yes
Voltage	120V	120V	120V
Size	14.5" x 8" x 15.5"	16" x 6" x 12"	11.5" x 14.25" x 11.5"
Availability (in-store, online)	Online	Online	Online
Price	$399	$399	$429
Warranty	2 years	1 year	1 year

The benefits to using a water oven over a circulator include having a secure lid to prevent evaporation, double-wall insulation to conserve energy while maintaining water temperature, optional racks to make cooking separate pouches or jars easier, and a lack of exposed moving parts to eliminate mineral deposits that may cause the unit to malfunction. These concerns with circulators are typically addressed with home hacks and added maintenance. Water ovens take less maintenance and tweaking, but they do take up more dedicated kitchen space. That said, most people find them more attractive, less of a hassle to maintain, and quieter in operation compared with circulators.

Other Necessary Equipment

Besides requiring a method to control and maintain the temperature of your water bath, there are a few other items you'll want on hand. Some are essential, some are nice to have, and some are just downright fun. Of course you'll need a way to seal your food and protect it from the water bath. For most meats and

DIY

If you're not ready to buy a dedicated sous vide machine, it's entirely possible to create your own rig at home. There are simple ways to test the waters (get it?) that include carefully monitoring the temperature of a pot on your stove, filling an insulated cooler with warm water, or plugging a slow cooker into a temperature controller. But if you're the more adventurous type, you can build an entire rig from parts you can buy online for about $50.

The most basic way to get started with sous vide is to fill a pot with water and set it on your stove. This will require you to maintain a consistent temperature using a thermometer and to increase or decrease the heat level as needed. Be sure to stir the pot occasionally to make sure the heat is being distributed evenly. It's certainly time-intensive, as most dishes take at least an hour to cook, but if you're up for the challenge, you'll yield results as good as those of any machine.

For those who don't want to stand by their stove all day, another tested method is the beer cooler hack. Fill an insulated cooler with water a few degrees hotter than you want to use for cooking, which will allow for temperature loss when you add the food to it. The cooler will keep the water bath at a stable temperature for a few hours, allowing you to cook steaks, chicken, and fish without a problem. The drawback is with higher temperatures and with longer cooks. Coolers will lose up to 1° per hour in the 140°F to 150°F range and are definitely not recommended for roasts that require 24 hours, or vegetables that need much higher temperatures to cook.

If you've already got a rice cooker or slow cooker in your kitchen arsenal, you may want to try it out as a water bath using on-the-market devices to precisely control the temperature. Codlo (right) and DorkFood both offer options that allow you to plug an analog cooker (meaning it has no digital on/off switches) into their units to maintain determined temps. The only drawback is that the size of your water bath is typically smaller, and there's no water circulation, meaning you can't guarantee even temperature distribution. Nonetheless, it is a very affordable entry into sous vide.

For the more industrious tinkerers and builders, there are many great tutorials online for how to build your own electrical unit complete with immersion heaters, aquarium water pumps, and PID temperature controllers. Instructables.com offers lots of tutorials to hack together parts you can source online, including the first-ever version of the Nomiku circulator. For a more elegant (and intensive) solution, check out SeattleFoodGeek.com for the model that launched the Sansaire.

for some vegetables, you'll want a heat source to finish the food, giving it a nice sear and flavor. And it's good to have a thermometer on hand both to make sure your circulator is correctly calibrated and to check the finished temperature of your dish.

BAGS

When it comes to bagging your food, you can easily get away with using sealable zip-top bags. They are BPA free, easily sourced, and very affordable. Many people even wash and reuse their bags. However, there are a few drawbacks. First, it's impossible to create a true vacuum seal with these bags. With air in the bag, you're not using the heat transfer properties of the cooking technique as efficiently as it was intended, and your food is more likely to float. Second, if you're cooking something for more than 24 hours, the likelihood that your seal will give out increases—I made this mistake myself once upon a time. The safer alternative is cooking in vacuum-seal bags (very affordable when bought by the roll, and readily available online). They do, of course, require a vacuum sealer.

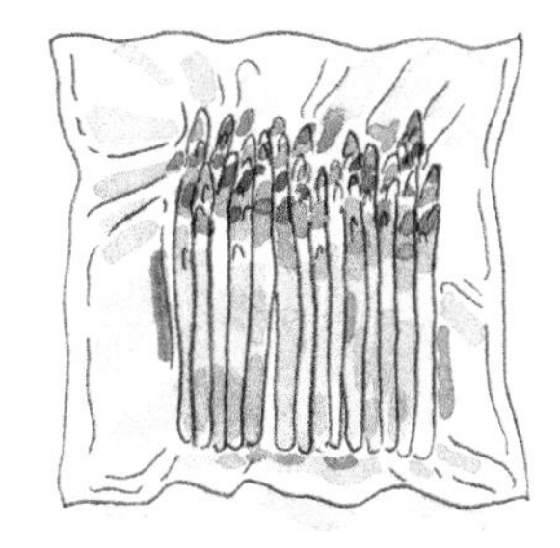

VACUUM SEALERS

Vacuum sealers come in a couple of varieties. Clamp-style external vacuum-sealing machines are the most common for home cooks. With these, you place the open end of your bag inside a tray in the machine, which then draws the air directly out of the bag through the open end. Because of this, you cannot truly vacuum seal liquids because they get sucked out of the bag and into the machine, which interferes with the seal. Also, you must use special bags that have texture in them to allow the air to move past the food and create the vacuum.

With a chamber vacuum-sealing machine, the entire bag is placed inside the chamber and a lid is brought down over it. The machine removes all of the air from within the chamber, and when the vacuum cycle is finished, the heat bar

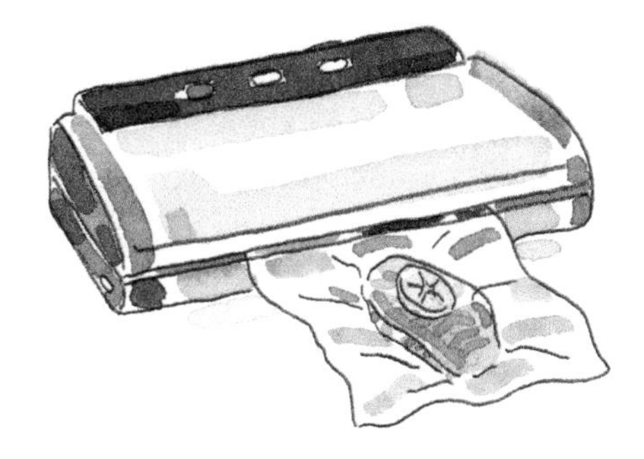

CURIOUS ABOUT COOKING WITH PLASTIC?

You may be concerned about the safety of cooking with plastic bags—and that concern is understandable. The major issues most have with plastics are phthalates, bisphenol A (BPA), polybrominated diphenyl ethers (PBDE), and tetrabromobisphenol A (TBBPA)—all of which alter hormone expression in humans and animals. These chemicals are known as endocrine-disrupting compounds (EDCs), and each affects different elements of hormone disruption (e.g., inducing estrogen-like activity, thyroid hormone homeostasis disruption, anti-androgens, and so on).

Most bags, including Ziploc and FoodSaver bags, contain polyethylene—not BPA or phthalates. Polyethylene is considered biologically inert, and scientists have been unable to detect any toxicity in animal tests (unlike plastics with BPA). It passes the Ames test (a test to see if a chemical causes DNA mutation) and other studies of damage to DNA. Additionally, you will never be using temperatures high enough (195°F) to soften the plastics, which is when chemical leaching would occur.

If you're concerned about the dangers of plastic despite this or feel it's not appropriate for your household, there are alternatives. Many foods can be prepared in vacuum-sealed glass jars (be sure to get lids that are BPA-free) and silicone bags (though some studies also pose concerns about the safety of even food-grade silicone). And never take risks with someone with a compromised immune system.

rises and applies the seal to the bag. Because the machine is never sucking on the open end of the bag, you can seal liquids in a bag. External vacuum sealers tend to be more compact and affordable, though the bags can be slightly pricier. Chamber sealers are larger and more expensive, but they allow you much more flexibility in what you can seal and tend to provide a more reliable seal.

THERMOMETERS

Thermometers are essential tools to ensure the correct internal temperature of your food. When you're cooking sous vide, thermometers have the added benefit of allowing you to double-check the temperature of your water bath to make sure your machine is correctly calibrated. I recommend an instant-read thermometer like the ThermoWorks Thermapen, which will give an accurate reading in only three seconds. For the more budget conscious, the Lavatools Thermowand works almost as quickly with no extra bells and whistles. Of course, any meat thermometer will do.

FIRE

When you're ready to take your food out of the water bath, it's time to give it a good sear. Sous vide can cook your meats expertly, but it will never get them to brown. What makes that seared crust so delicious is the Maillard reaction: a chemical reaction between amino acids and sugars that creates hundreds of new molecules responsible for the characteristic smells of roasting, baking, and frying.

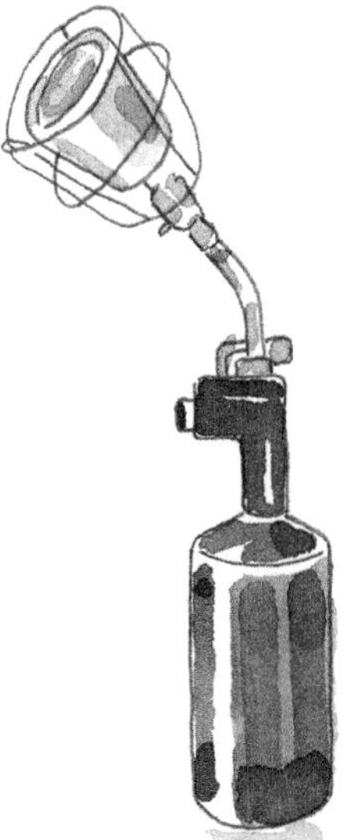

In order to achieve this reaction, we need high heat! The simplest solution is a cast iron pan. Cast iron has great heat retention and can get really hot, but a stainless steel frying pan (which typically contains an aluminum or sometimes a copper core) will also work. The key is to get the oil really hot before adding the meat, so use vegetable oil, which has a high smoke point. Be sure to pat the meat dry before searing or you'll end up steaming it rather than searing. When searing, the oil is less of a cooking medium and more of a way to get uniform surface contact between the meat and the pan. This produces nice, even browning and prevents some spots from burning while other spots remain pale. As it's heating, swirl the oil around to get a thin coating over the bottom of the pan. When the oil starts to shimmer and smoke just slightly, you're ready to add the meat.

Those who want more flair in their cooking should invest in a torch. The best torches for searing sous vide are ones designed for "industrial" uses such as soldering copper pipes, brazing, and light welding. These can reach temperatures greater than 3,500°F, which will provide enough heat to sear a typical sous vide dish in 2 to 3 minutes. A crème brûlée torch is not going to cut it. A BenzOmatic trigger-start torch is your go-to flame thrower when you want to impress. It's affordable, easy to use, and very impressive.

The Searzall has recently emerged as a sous vide–specific add-on to the BenzOmatic, promising a more even sear and less "torch taste" by using a shorter flame. The key here is that with the Searzall, the flame from the torch gets spread out into a more even, radiant heat, causing it to act as a handheld broiler. My experience with it was that it didn't provide any discernable difference in flavor, but it did make me concerned I was going to set my kitchen on fire. The flame is widely dispersed, which is intended to create a more even sear, but the amount of flame that burns up around the edges of the device is a concern. It's tricky to calibrate and requires burning off a layer of insulation before use. It's an impressive and showy piece of equipment, but use at your own risk.

Finally, deep fryers, smokers, and grills are also excellent ways to finish your meals if you have access to them. There's nothing more impressive than bringing out a rack of ribs that have been cooking in a water bath for 24 hours and dropping them on a grill for the best barbecue ever.

JARS

Canning jars are essential for cooking custards, yogurt, and other nonsolid foods. Technically, you could cook them in bags, but this way they come out of the bath ready to serve! Jars are also perfect for dealing with liquids for infusions, pickles, and confits. Be sure to use jars that are intended for high heats so they don't break when introduced to the water bath.

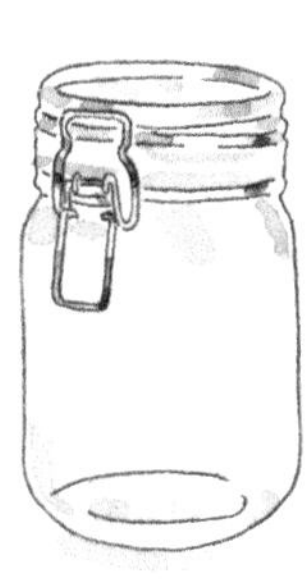

TONGS

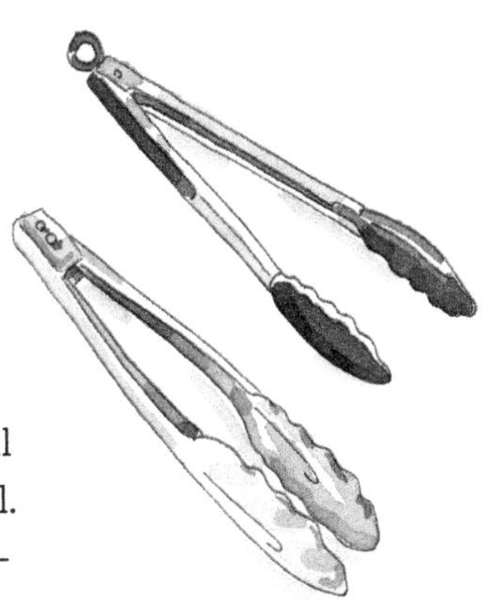

A good set of long, strong tongs is going to serve you well when it's time to take your cooked food out of its hot bath. I have two sets of tongs for this purpose—one set that's stainless steel and one that has silicone tips. The silicone tips are great for gripping bags and jars without worrying that they'll slip or tear, and the stainless set is ideal for turning big cuts of meat in a cast iron pan or on a hot grill. Always use the silicone tipped pair whenever you use a non-stick frying pan so you don't scratch the cooking surface.

BINDER CLIPS

This may sound funny, but some large, strong binder clips are super handy when it comes to keeping your food submerged. Lower your bags into the bath vertically, and clip them to the side of the container to prevent them from floating.

WEIGHTS

In case the binder clips still don't keep your bags submerged, you may need to place something on top of them to weigh them down. I've found the simplest solution to be a plate or bowl. I prefer to use glass so I can still see the food below the dish, but this isn't entirely necessary.

Fun Fact: Close the lids on the jars just tight enough for the lid to stay on so any air will be able to escape from the jars when you submerge them in water. If you close them too tightly, the trapped air will press against the glass and could crack or break your jars.

COVER

When cooking for long periods of time, water evaporation is a concern. Instead of keeping a close eye on your water bath, invest in a cover for your cooking vessel. This could be as simple as plastic wrap or as specific as a lid with a hole you cut out to suit the dimensions of your circulator. You can also purchase BPA-free plastic balls or Ping-Pong balls to float on the surface of your water bath. These are convenient because they'll fit any size container and prevent evaporation and heat loss during cooking by creating a floating barrier.

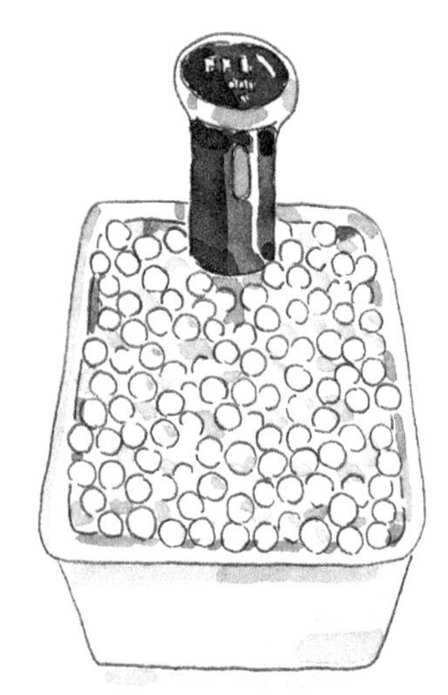

Sous vide has the power to transform even a novice cook into someone who turns out amazing meals every time. With sous vide, you'll never again overcook a steak, undercook pork chops, or turn scallops into little discs of rubber. Now that your kitchen is ready to go, it's time to dive into sous vide cooking basics. By learning a few simple principles, you'll be able to apply these methods to any meal you want to make and elevate your status to master home chef.

Fun Fact: "Sous vide" actually only describes the process of vacuum sealing the foods. When we cook, we're using low-temperature water baths, and there just isn't a clever term for that yet.

2

BEFORE THE BATH

SOUS VIDE BASICS

Consider this chapter your complete primer on cooking absolutely anything sous vide. Here, we'll cover everything you need to know about how to prep, season, and cook your food; avoid missteps; and get yourself set up to prepare the best meals of your life. Once you learn a few simple basics, you'll be ready to mix and match recipes, explore new foods and flavor combinations, and experiment.

Three Steps to Sous Vide

Just as you should always read through an entire recipe before you start cooking, it's good to know every step you plan to take before embarking on your sous vide journey. Before you begin, map out which elements of your meal you plan to cook sous vide and which you'll prepare using other methods. Get familiar with the appropriate time and temperature you'll need for what you're cooking. Then decide how you're going to bag up, seal, and finish your meal. Once you have this roadmap laid out, you're ready to go.

FOOD PREP

Truth be told, most meats do not need to be fully seasoned before they cook. Since salt is the only seasoning that can actually penetrate muscle tissue, there's really no benefit to adding anything else to the bag. In fact, some flavors like raw garlic and onion can actually be detrimental to the finished product because the water bath for meats is not hot enough to actually cook them (and mellow their pungency). If you truly want to impart flavors into your meats prior to cooking, you can brine or marinate them beforehand. That said, there's usually nothing wrong with adding your favorite seasonings and spice rubs, which can actually make the process more fun and flavor the liquid that exudes during cooking. You can even separate your meats into individual bags and use different seasonings on each—perfect for a family of picky eaters or when preparing a week's worth of meals.

If you salt food before cooking it sous vide, holding the food at low temperatures for a long time may inadvertently cure the food, resulting in a dense, pickled texture. This typically isn't a problem for food you're going to cook and serve immediately. But if you plan to refrigerate or freeze your meat after cooking it, the salt can continue to penetrate the muscle tissue, so consider adding your seasonings right before you serve.

Meats that are vacuum sealed will take on the shape of the bag, which can be undesirable when serving. It's easy to trim the meats before you serve them to make them more appealing, but some people suggest other methods prior to placing them in the water bath. Food can be cooked from frozen, which helps prevent it from taking on the vacuumed shape. And you can pre-sear softer meats to help them firm up and hold their shape better.

MIX AND MATCH

Most people who cook sous vide will employ a variety of techniques to complete their meal. You may choose to cook your proteins using sous vide and prepare your vegetables separately since they usually take less time and cook at different temperatures than meats. You could also opt to start by cooking your vegetables sous vide, then lowering the temperature of the water bath to cook your meat, letting the vegetables hang out in the bath and stay warm. In fact, no matter how you decide to cook your sides, you can always add them to a bag and drop them in the water bath with the meat to keep them warm before serving.

If you have access to a smoker, pre-smoking your meats before sous vide lends them a wonderful depth of flavor that continues to penetrate the meat while it cooks. And, as we've already discussed (page 23), there are several methods you can use to finish cooking your meats to give them that great sear. Whatever method you choose, make sure to have a plan before you get started, and you'll be on the path to success.

SEAL AND SET

Once you've prepped your food, it's time to get it bagged up. Vacuum sealing allows for the most efficient heat transfer from the water to the food. The low-tech method to do this is called the "displacement method." Simply put, add your food to a zip-top bag, and close the seal most of the way, leaving only the corner open. Slowly lower the bag into water, which will cause the air to be forced out the top. When almost all the bag has been submerged, but before any water can find its way into the bag, close the seal completely. You'll still have a little air in the bag, but for most purposes, this works well enough. If you are concerned about the seal leaking, double-bag your food.

The simplest way to create a completely air-free environment in your bags, of course, is with a vacuum sealer. The more affordable clamp-style model is great for most purposes, but you need to be careful to make sure that it creates a complete seal. If for any reason it does not or you notice too much air left in the bag, you can always open and reseal it. What's great with clamp-style sealers is that many have the option of a gentle vacuum setting, which is ideal for softer meats and fish.

One drawback when using a clamp-style vacuum sealer is that liquids can get sucked out of the bag and keep it from sealing properly. To avoid this, always choose the "moist" setting on the vacuum sealer. Better yet, freeze any seasoning liquids before placing them in the bag for vacuum packing. I recommend filling an ice cube tray with olive oil so you always have some cubes on hand when cooking sous vide. (Avoid using extra virgin olive oil, as some people find it adds an "off" flavor when held at low temperatures for a long time.) Chamber sealers can actually vacuum seal liquids, and zip-top bags used with the displacement method described previously provide another easy solution for recipes requiring liquid in the bag.

However you bag your food, don't overcrowd the bags, and (with solid foods) keep the food in a single layer so it cooks evenly.

Prepare your water bath by making sure you have enough room for the water to circulate freely around the food and enough volume to completely submerge the bag(s). Always make sure the water bath has been heated to the temperature you want before you add the bags and begin the cook. You can expect the temperature to drop when you add the bags in to cook, but it will only take a few minutes to regulate.

LET'S GET COOKING!

Now that everything's seasoned and sealed, it's time to cook. Carefully lower your food into the water bath, making sure it is completely submerged. Be very careful—although the water isn't hot enough to burn you, it's still pretty hot to the touch! Use those tongs with silicone tips to help lower and remove your food from the water bath. One study has shown that cooking your bags vertically allows for better heat transfer, but if that's not possible, make sure there's even circulation around all sides of the bags for best results.

Fun Fact: Fill the water bath with hot water rather than cold to expend less energy and reach your cooking temperatures more quickly.

THE IMPORTANCE OF TEMPERATURE CONTROL

As we've learned, what sets sous vide apart from traditional cooking methods is the ability to precisely control the temperature at which our food is being cooked. While it might be tempting to stick with cheaper DIY solutions, there are some obvious downfalls. First, you need to keep a keen eye on the water bath temperature to keep it from getting too high and overcooking foods or dropping too low, which could be a safety hazard.

Precise temperature control allows for greater control over doneness than does traditional cooking methods. Food can be pasteurized at temperatures as low as 129°F so that it does not have to be cooked well-done to be safe, and tough cuts of meat (which were traditionally braised to make them tender) can be made tender at medium or medium-rare temperatures. All of this depends on equipment that will maintain reliable, accurate temperatures.

The major concern with cooking food at low temperatures is propagation of bacteria. You may have heard of the "danger zone" of 40°F to 140°F, where bacteria supposedly thrive. The biggest misconception about bacteria and the danger zone is that any food in the temperature range is not safe and as soon as you move above 140°F the food instantly becomes safe. The truth is that food safety is a function of both temperature and time; most bacteria can be killed as effectively by long periods at lower temperatures as by high temperatures for a short time. What this means for us is that we can achieve perfectly safe meats at much lower temperatures when we cook them for longer periods of time.

Temperature also affects the proteins and connective tissues in the meat we cook. Muscle fibers start to shrink and connective tissues contract at high temperatures, causing the meat to expel liquid and become tough. By cooking at low temperatures for long periods of time, we can attain meats that stay juicier and more tender than is possible with traditional methods.

One advantage to sous vide cooking is that you can cook your food and store it in the sealed pouches in the refrigerator or freezer, ready to be reheated and served when necessary. As long as you reheat it below the target cooking temperature, your food will be perfectly cooked. However, when using this process, you need to be extra careful with food contamination and follow all necessary procedures to ensure the food is safe to eat. As soon as the food is done cooking, drop the bags in an ice bath to immediately stop the cooking process and drop the temperature below the temperature zone in which bacteria can grow and thrive (40°F to 140°F). It's not enough just to put the food directly in the refrigerator, since it will take too long to reach safe temperatures and risks raising the temperatures of surrounding foods. When you're ready to eat, you can reheat it in a sous vide water bath at or below the temperature it was cooked in and finished using whatever method you like.

If you're planning to serve it immediately, there's no safety reason to shock your food, but in some cases, you may want to anyway. With scallops, for example, a quick ice bath prior to searing can prevent them from dumping their juices as soon as they hit the hot pan. But for the most part, you're ready to go. Carefully remove the bags from the water bath with your tongs. Open the seal using a sharp knife or kitchen scissors. Remove the food from the bag to a plate lined with paper towels. You can reserve any juices from the bag for a pan sauce (directions follow) or discard them. Pat the food dry before searing to allow for the best crust. If you're finishing meats in a hot pan, it's great to toss in the vegetables afterward to pick up any remaining little delicious brown bits (fond) stuck to the pan. Doing so adds extra flavor to your side dishes and pulls the entire meal together.

Fun Fact: When cooking with canning jars, always be sure they are clean and sterilized. Prepare jars by sterilizing with boiling water or in the dishwasher.

USING LEFTOVER LIQUIDS

After cooking meats for a long period of time, there may be a lot of liquid left in the bag. All cooking methods pull water out of meats, but other methods can hide it better than sous vide. With roasting or grilling, the juices evaporate or drip away. What's great about sous vide is that you retain all of the flavorful juices for a really fantastic sauce.

What you need to know before adding them straight to your saucepan is that since the meats have been cooked at such low temperatures, the proteins are going to curdle when you heat them any further—which is exactly what you'll do to remove them from the liquid.

1. In a saucepan or microwave-safe container, heat the liquid until boiling.
2. Next, line a sieve with damp paper towels (or cheesecloth) and filter the juices.
3. Discard the solids and chill the liquid until the fat separates.
4. Once the liquid is chilled, you can remove the fat and keep the clarified liquid.

If there were any seasonings in the bag, the clarified liquid will retain all of those flavors. You can brown the meat and then deglaze the pan with the strained juices to make a quick sauce.

For a fancier sauce:

1. Sauté chopped garlic with a few thyme sprigs and a bay leaf for 1 minute in any fat left in the pan from browning the meat.
2. Deglaze the pan with wine and add the strained juices.
3. Reduce over medium heat for a few minutes and then strain. Whisk in some small pieces of cold butter and serve.

EGGS, SOUS VIDE STYLE

Eggs are the perfect introduction to sous vide cooking. Because the whites and yolks exhibit different qualities at different temperatures, they provide you with the perfect playground to experiment up to 0.1°. Plus, they come with their own built-in watertight shells, so you don't have to fuss with bags. With sous vide, you can perfect the poach; create rich, fudge-like yolks; and master hard-boiled with no hint of chalkiness.

SOFTBOILED EGGS

YIELD: 1 TO 12 EGGS / ACTIVE TIME: 5 MINUTES / TOTAL TIME: 45 MINUTES

1 to 12 eggs

1 Preheat the water bath to 143°F.

2 Add the eggs to the water bath and cook for 45 minutes.

3 Remove from the water and serve immediately. To store the eggs, cool immediately in an ice bath and store in the refrigerator for up to 3 days. Reheat in a water bath at 135°F for 30 minutes to serve.

HARDBOILED EGGS

YIELD: 1 TO 12 EGGS / ACTIVE TIME: 5 MINS / TOTAL TIME: 45 MINUTES

1 to 12 eggs

1 Preheat the water bath to 165°F.

2 Add the eggs to the water bath and cook for 45 minutes.

3 Serve immediately or chill in an ice bath if desired. To store the eggs, cool immediately in an ice bath and store in the refrigerator for up to 3 days.

SOFT-POACHED EGGS

YIELD: 1 TO 12 EGGS / ACTIVE TIME: 5 MINUTES / TOTAL TIME: 50 MINUTES

1 to 12 eggs

1 Preheat the water bath to 143°F.

2 Add the eggs to the water bath and cook for 45 minutes.

3 Remove from the water and allow to cool slightly.

4 Bring a medium pot of water to a bare simmer.

5 Taking one egg at a time, carefully crack the egg close to one of its ends.

6 Hold a perforated spoon over a small bowl and slip the egg out of its shell onto the spoon, allowing the excess loose white to drain into the bowl. Discard the excess white and place the egg in a larger bowl. Repeat with the remaining eggs.

7 Once all the eggs are drained, carefully slip the eggs into the simmering water, gently swirling the water occasionally to prevent the eggs from sticking to the bottom.

8 Cook until the outer whites are just set, about 1 minute.

9 Remove the eggs from the water with the (cleaned) perforated spoon and serve immediately.

Pro Tip: *You can actually store poached eggs in cold water in the refrigerator for up to 3 days. To reheat, place the eggs in a bowl of hot water for a few minutes until warm. Perfect when prepping brunch for a crowd!*

Troubleshooting

Every so often, you may encounter some unexpected hiccups when cooking sous vide. Most can be solved with some creative solutions. What follows is a list of common sous vide problems and how to fix them.

HARDWARE ISSUES

If you ever have a hardware problem, consult the manufacturer's website and contact their customer service as needed. I've been in touch with every main manufacturer of sous vide equipment to date and can report excellent support on all fronts.

COOKING DIFFERENT SIZE CUTS OF MEAT

With traditional cooking methods, it can be tricky to figure out how long to cook different size cuts of meat to get them all to come out perfectly. Fortunately, this problem is immediately solved by cooking sous vide. Since you're already cooking the meat at its ideal temperature, there's no need to worry about overcooking one and undercooking another based on its size. Just be sure the larger cut is heated all the way through using the time and temperature tables provided, and you're good to go. In almost all cases, the smaller cut will not suffer from staying in the water bath.

FLOATING BAGS

The most common difficulty faced when cooking in a water bath is to keep the food from floating. The simplest solution to this is to use large binder clips and fasten the bag to the side of your vessel. Most of the time the weight of the food in the bag will be enough to keep it submerged with this method. Sometimes adding a little oil to the bag will also help keep it submerged. If this still doesn't work, place a weight on top of the bag, like a glass plate or bowl.

WATER LEVELS

When cooking foods for 12 hours or more, it's important to keep an eye on your water levels. You don't want your food to be even partially exposed to air, allowing the temperature of the food to drop below that of the water bath and enter the danger zone. One solution to having to constantly top up the water bath is to cover the cooking vessel with plastic wrap. Foil also works, but since foil conducts heat it's not a very energy-efficient solution. You can also purchase insulated floating balls, which are marketed specifically for preventing evaporation and heat loss.

LEAKS

When you've invested the time into a three-day cook, the last thing you want is to come into your kitchen to discover that the bag has leaked. While you're probably not in trouble from a safety viewpoint since your food is still submerged at the appropriate temperatures, your food may now be waterlogged and unsavory. If you catch the leak in time, you can still rescue your food—just take it out, dry it off, and reseal. You may need to replace the water in the bath if too much liquid has leaked out into it. Seal leaks occur more often with zip-top bags and can be prevented by double-bagging. Make sure to use the displacement method each time to remove as much air as possible.

Now that you know the basics of how to cook sous vide, you're ready to start preparing the most amazing meals of your life.

PART 2

THE RECIPES

3

BEEF AND LAMB

Sous vide is an excellent way to get consistent roasts and chops every time, but the technique truly shines with steak. If you've ever been to a high-end steakhouse, you've had a sous vide steak—perfectly cooked from edge to edge with a thin, crisp crust. Sure, you can cook great steaks over a hot grill or stove, but getting a perfectly medium-rare center every time is no easy feat. One minute too little or too much means underdone or overcooked, and varying cuts and thicknesses can wreak havoc on your results. With sous vide, you're in no rush. You can even cook steaks to different temperatures to suit the tastes of various guests and still have them all come out perfectly and at the same time.

Tender Cuts: Tenderloin, Rib Eye, Strip, Sirloin

YIELD: 1 TO 12 STEAKS / ACTIVE TIME: 10 MINUTES / TOTAL TIME: 1 HOUR, 10 MINUTES

Because tenderloin has less fat and intramuscular fat, it can be cooked at lower temperatures and for shorter times than other cuts. Highly marbled cuts like rib eye and strip should be cooked a few degrees higher since their fat helps keep them moist while delivering plenty of flavor. Use the times and temperatures on page 45 as a guide to find your perfect steak. Cooking temperatures depend on your preferred doneness, while times can vary between 1 hour (or 45 minutes for the less fatty tenderloin) at the low end and 4 hours at max. Rare steaks should not cook for more than 2 hours.

1 (1½- to 2-inch) tenderloin fillet, rib eye, or strip steak per person
Salt
Freshly ground black pepper
1 tablespoon vegetable oil
Favorite spice blend (optional, to replace salt and pepper)
Rosemary, thyme, or other aromatics (optional)—1 sprig per steak
Garlic cloves (optional)
1 tablespoon butter

1 Preheat the water bath to the desired final temperature.

2 Season the steak generously with salt and pepper or your favorite rub.

3 To bag the steak, start by folding the top of a vacuum-seal or zip-top bag back over itself to form a cuff. (If cooking more than 1 steak, for best results, do not bag more than 1 large or 2 small steaks together in one bag.)

4 Add the steak and herbs (if using). Uncuff and seal the bag using either a vacuum sealer or the displacement method.

5 Place the bag in the water bath and cook for 1 hour. (The steaks can remain in the water bath for up to 4 hours, but the ideal cooking time is 1 hour.)

6 Remove the steaks from the bag, place them on a paper towel–lined plate, and pat dry on both sides.

7 Heat the oil in a large cast iron or stainless steel skillet over high heat until the oil is almost smoking. Add the steaks. Add aromatics like whole thyme and rosemary sprigs or crushed whole garlic cloves (if using).

8 Cook for 15 to 30 seconds and flip the steaks with your tongs. Repeat, flipping the steaks every 15 to 30 seconds until they have developed a nice brown sear, about 90 seconds total. Holding the steaks with the tongs, turn them vertically to sear the edges all around. Work quickly to avoid overcooking the them.

9 Add the butter to the skillet and let it melt, tossing the melting butter onto the top of the steaks, for 30 seconds.

10 Alternatively, sear the steaks on a grill or with a torch, making sure to brown all sides.

11 Sprinkle with freshly ground pepper and sea salt and serve.

Tip: *You can cook steaks straight from the freezer. For steaks 1 to 2 inches thick, allow an extra hour for the steak to fully thaw in the hot water bath before you begin timing it for doneness.*

RARE	125°F to 128°F
MEDIUM-RARE	129°F to 134°F
MEDIUM	135°F to 144°F
MEDIUM-WELL	145°F to 155°F
WELL DONE	155°F and up

Tough Cuts: Flank, Brisket

YIELD: 2 TO 4 POUNDS / ACTIVE TIME: 10 MINUTES / TOTAL TIME: 8 TO 48 HOURS, DEPENDING ON SIZE OF CUT AND DESIRED DONENESS

Tough cuts of meat require much longer cooking times than do tender meats to break down the connective tissues, but the results are well worth the wait. Sous vide is ideal for tough cuts because you can cook an entire roast to medium-rare without losing any of the juiciness. Cook a thinner, 2-pound, flank steak for at least 8 hours and up to 24 hours. For a thick, 4-pound, brisket, cook for 24 hours for steaks or sandwiches. For tender, fall-apart shredded beef, cook the brisket for up to 36 hours.

2 to 4 pounds brisket or flank steak
2 tablespoons vegetable oil
Salt
Freshly ground black pepper
2 tablespoons Steakhouse or Smoky Spice Rub per pound (optional; see pages 214 and 215)

1 Preheat the water bath to the desired final temperature.

2 Place the meat on a paper towel–lined plate and pat dry on all sides. Heat the oil in a large cast iron or stainless steel skillet over high heat until the oil is almost smoking. Add the meat and turn to sear on all sides, about 2 minutes per side.

3 Transfer the meat to a plate and season all over with salt and pepper or a spice rub (if using).

4 To bag the meat, start by folding the top of a vacuum-seal bag back over itself to form a cuff. For best results, do not use zip-top style bags. Add the meat. Uncuff and seal using a vacuum sealer.

5 Place the bag in the water bath and cook for at least 8 hours and up to 48 hours, depending on the size of the cut and desired tenderness.

6 Remove the meat from the bag, place it on a paper towel–lined plate, and pat dry on all sides.

7 If you want to recrisp the crust, you can sear it again. Use tongs to turn the meat in the hot pan. This step is optional. Flank and brisket are great served alone or can be used in a variety of dishes.

Tip: *Because these cuts are generally larger and can be hard to sear after prolonged cooking times, I recommend pre-searing the meat.*

MEDIUM-RARE	131°F	8 to 36 hours
MEDIUM	140°F	8 to 36 hours
TRADITIONAL OR WELL-DONE	160°F	8 to 48 hours

48-Hour Chuck Roast

YIELD: 3 TO 5 POUNDS / ACTIVE TIME: 10 MINUTES / TOTAL TIME: 24 TO 72 HOURS, DEPENDING ON DESIRED TEXTURE

Not only does sous vide turn a chuck roast into mouthwatering morsels of tenderness, it's also super flexible because you don't have to decide how to season it until it's done cooking. By seasoning minimally before the roast goes into the water bath, you can repurpose your roast for a variety of dishes: pot roast, pulled barbecue beef, with chimichurri, and more.

3- to 5-pound chuck roast
2 tablespoons vegetable oil
¾ cup Steakhouse or Sweet and Spicy Rub (pages 214 and 216)
A few dashes liquid smoke (optional)

1 Preheat the water bath to 140°F.

2 Pre-sear the brisket, because it's easier to handle at this stage. Heat 1 tablespoon of the oil in a large cast iron or stainless steel skillet over high heat until the oil is almost smoking. Add the brisket and turn to sear on all sides, about 2 minutes per side.

3 Remove from the heat and season liberally with spice rub on all sides.

4 To bag the brisket, start by folding the top of a vacuum-seal or zip-top bag back over itself to form a cuff. Add the brisket and the liquid smoke (if using). Uncuff and seal the bag using either a vacuum sealer or the displacement method. (If using zip-top bags, seal in a second bag as well.)

5 Place the bag in the water bath and cook for 24 to 72 hours, depending on desired tenderness.

6 Remove the brisket from the bag and place on a paper towel–lined plate and pat dry on all sides. Discard the juices or reserve them to create a sauce (see the section Using Leftover Liquids, page 35). Pat dry and trim away the fat.

7 Heat 1 tablespoon of the oil in a large cast iron or stainless steel skillet over high heat, until the oil is almost smoking. Re-sear the brisket on all sides to create a crust, about 2 minutes per side.

8 Slice or shred and serve as desired.

Tip: *For longer cooks, carefully rotate the bag every few hours to help flavor the food evenly.*

SOFT AND TENDER, SLICEABLE	150°F	24 hours
JUICY AND SHREDDABLE	140°F	48 hours
MELT-IN-YOUR-MOUTH TENDER	134°F	72 hours

Melt-in-Your-Mouth Brisket

YIELD: 4 POUNDS / ACTIVE TIME: 10 MINUTES / TOTAL TIME: 24 TO 48 HOURS, DEPENDING ON DESIRED DONENESS

Brisket is a big, tough cut of meat with lots of collagen and connective tissues that need long, low heat to break down. Traditionally, a smoker does the job, but sous vide provides a perfect alternative. You can get the low and slow cook nailed from your own countertop and add that hit of smoke with liquid smoke and smoked paprika. A hot grill is perfect for finishing this barbecue classic, but a hot cast iron pan works too.

1 or 2 tablespoons vegetable oil
¾ cup Smoky Spice Rub or Memphis Dust (pages 215 and 216)
A few dashes liquid smoke (optional)

Challenge: *For a more nutrient-dense meal, substitute the brisket with beef heart! It will cook in the same time and produce an even richer flavor.*

1 Preheat the water bath to the desired final temperature.

2 Place the meat on a paper towel–lined plate and pat it dry on all sides. Heat 1 tablespoon of oil in a large cast iron or stainless steel skillet over high heat until the oil is almost smoking.

3 Add the brisket and turn to sear on all sides, about 2 minutes per side. Transfer to a plate and season liberally with a spice rub on all sides.

4 To bag the brisket, start by folding the top of a vacuum-seal or zip-top bag back over itself to form a cuff. Add the brisket and the liquid smoke (if using) to the bag. Uncuff and seal the bag using a vacuum sealer or the displacement method. If using zip-top bags, I recommend double-bagging.

5 Place the bag in the water bath and cook for 24 to 48 hours, depending on desired tenderness.

6 Remove the brisket from the bag, place on a paper towel–lined plate, and pat dry on all sides. Discard the juices or reserve them to use in stock or a pan sauce (see the section Using Leftover Liquids, page 35). Trim away the fat.

7 Re-sear on a hot grill, or heat 1 tablespoon of oil in a large cast iron or stainless steel skillet over high heat until the oil is almost smoking. Add the brisket and turn to sear on all sides, about 2 minutes per side. Slice and serve.

MEDIUM-RARE	130°F	24 to 36 hours
MEDIUM	140°F	24 to 36 hours
MEDIUM-WELL	150°F	24 to 48 hours
WELL-DONE	160°F	24 to 48 hours

The Juiciest Burgers

YIELD: 4 BURGERS / ACTIVE TIME: 10 MINUTES / TOTAL TIME: 50 MINUTES TO 4 HOURS, DEPENDING ON DESIRED DONENESS

Why cook a burger sous vide? Why not just drop it in a hot pan or on the grill and call it a day? That's fine if you're cooking skinny little burgers, but for a fat, juicy half-pounder, you're going to want to pull out the full arsenal. Cooking a burger sous vide offers advantages in that it's totally foolproof, it can cook your burger to a perfect medium-rare all the way through, and yet it still allows for the perfect sear.

2 pounds freshly ground beef
Salt
Freshly ground black pepper
Favorite spice blend (optional, to replace salt and pepper)
2 tablespoons vegetable oil, divided
4 hamburger buns
4 slices cheese (optional)
Ketchup, mustard, mayo, pickles, lettuce, onions, and other preferred burger condiments

Note: *Temperatures are lower than typical for cooking beef because burgers are thinner and less dense than steaks, meaning that they transfer heat more quickly. When seared, they'll gain a good 10°F, so a burger cooked to 120°F sous vide will actually be at around 130°F by the time it hits the bun.*

1. Preheat the water bath to the desired final temperature.
2. Divide the meat into four 8-ounce portions and season liberally with salt and pepper or your favorite spice blend (if using).
3. Gently pat the beef together to form patties. Do not overhandle or compress the beef.
4. Seal the patties individually in zip-top bags using the displacement method to avoid squishing them. Gently lower the bags into the water bath. Clip them to the edge if they have trouble staying submerged.
5. Cook for at least 40 minutes. Remove the burgers from the bags, transfer them to a paper–towel lined plate, and gently pat dry.

6 Heat 1 tablespoon of oil in a cast iron or stainless steel skillet over high heat until the oil just begins to smoke. Add the buns facedown and heat until lightly toasted, about 45 seconds. Set aside.

7 Add the remaining 1 tablespoon of oil to the skillet and add the patties. Cook until well browned on the first side, 45 seconds to 1 minute.

8 Flip the patties and add the cheese (if using). Cook until the second side is browned and the cheese is melted, 45 seconds to 1 minute longer.

9 Transfer the burgers to the prepared buns, top with condiments as desired, and serve immediately.

A note on safety: *Most bacteria live on the surface of meat, but when meat is ground, they can be found throughout the meat. At 130°F, it takes 2 hours to safely pasteurize beef, while at 140°F, it takes only 12 minutes. Remember, these time frames begin once the center of the burger reaches pasteurization temperature, so it's a good idea to add an extra half hour to those times for any burger you plan on pasteurizing. Pasteurization cannot safely take place lower than 130°F, so for this reason, it is strongly recommended to freshly grind beef for sous-vide burgers you plan on serving rare to medium-rare. If you can't do this at home, you can ask your butcher to do it for you.*

RARE	120°F to 124°F	40 minutes to 2½ hours
MEDIUM-RARE	125°F to 129°F	40 minutes to 2½ hours
MEDIUM	130°F to 137°F	40 minutes to 4 hours
MEDIUM-WELL	138°F to 144°F	40 minutes to 4 hours

Beef Short Ribs

YIELD: 4 SERVINGS / ACTIVE TIME: 10 MINUTES / TOTAL TIME: 12 TO 72 HOURS, DEPENDING ON DESIRED TEXTURE

Short ribs are one of those cuts that undergo miraculous transformations depending on time and temperature. This doesn't mean that longer and lower is necessarily better (though it's my preference); it just means that you can attain various results to suit your taste from this simple, inexpensive cut of meat. I encourage you to experiment to determine the cooking time and temperature you prefer. Once you've found your favorite, make sure to make a note of it for the future!

4 pounds (2 slabs, or about 8 ribs) beef short ribs
1 tablespoon vegetable oil
4 thyme sprigs (optional)
2 rosemary sprigs (optional)
2 whole garlic cloves, crushed (optional)
1 tablespoon butter

1 Preheat the water bath to the desired final temperature. Do not season the ribs if cooking them for more than 24 hours.

2 To bag the ribs, start by folding the top of a vacuum-seal or zip-top bag back over itself to form a cuff. Add the ribs. Uncuff and seal the bag using either a vacuum sealer or the displacement method. Do not crowd the bags; you will probably need at least two bags.

3 Place the bags in the water bath and cook for 12 to 72 hours, depending on desired tenderness.

4 Remove the ribs from the bags, place on a paper towel–lined plate, and pat dry on both sides. Discard the juices or reserve to create a sauce (see the section Using Leftover Liquids, page 35).

5 Remove the bones from the ribs and trim away the fat. At this point, the meat can be reserved and used for a sauce or ragu. For immediate serving, sear the meat.

6 To sear, heat the oil in a large cast iron or stainless steel skillet over high heat until the oil is almost smoking. Add the ribs, and add the thyme, rosemary, and/or garlic (if using). Cook for 45 seconds, then flip the ribs.

7 Add the butter to the skillet and cook for an additional 30 seconds.

TRADITIONAL BRAISE STYLE	185°F	12 to 24 hours
SOFT AND TENDER, SLICEABLE	158°F	16 to 24 hours
STEAK-LIKE, SUCCULENT	144°F	24 to 48 hours
VERY TENDER, JUICY, AND ROBUST	129°F	48 to 72 hours

Barbecued Beef Sandwiches

YIELD: 8 SANDWICHES / ACTIVE TIME: 10 MINUTES / TOTAL TIME: 10 MINUTES (PLUS UP TO 48 HOURS FOR THE BEEF)

Barbecued sandwiches are a staple of summer, but you don't need to slave over a smoker or grill to get that great smoky taste. You can prepare this recipe using either chopped brisket or shredded chuck roast. Just mix with your favorite sauce, or try the barbecue sauce on page 212, and top with coleslaw and pickles.

FOR THE SANDWICHES

8 hamburger buns
1 tablespoon vegetable oil (optional)
6 cups of beef from Brisket (page 50) or Chuck Roast (page 48), chopped or shredded
Sweet and Spicy Barbecue Sauce (page 212)
Sliced dill pickles

FOR THE COLESLAW

1 carrot, peeled and cut into thin matchsticks
¼ head green cabbage, thinly sliced
2 tablespoons white vinegar
2 tablespoons mayonnaise
½ teaspoon whole caraway seeds
½ teaspoon celery seeds
Salt
Freshly ground black pepper

Fun Fact: *You can reheat leftovers using sous vide by rebagging and reheating at or below the temperature they were cooked at.*

TO MAKE THE SANDWICHES

1 Toast the buns under the broiler or in a toaster, or heat the oil in a large skillet over medium-high heat until the oil shimmers. Add the buns facedown and heat until lightly toasted, about 1 minute.

2 Set aside.

3 Toss the chopped or shredded meat with the barbecue sauce.

4 Top the toasted buns with meat, pickles, and coleslaw and serve.

TO MAKE THE COLESLAW

In a large bowl, combine the carrot and cabbage. In a small bowl, stir together the vinegar, mayonnaise, caraway seeds, and celery seeds. Pour over the vegetables and stir to coat. Season with salt and pepper. Let stand for 15 minutes before serving.

Garlic-Rosemary Lamb Chops

YIELD: 4 SERVINGS / ACTIVE TIME: 10 MINUTES / TOTAL TIME: 2 TO 4 HOURS

Garlic and rosemary were made for lamb. Serve these delicate chops with fresh green peas and a salad for a light and flavorful springtime meal, or with Creamy Garlic Mashed Potatoes (page 154) for something heartier. This dish comes together with absolutely minimal effort, so it's perfect for a weekend dinner.

2 pounds lamb rib or loin chops (2 to 3 chops per person)
2 cloves garlic, minced
Salt
Freshly ground black pepper
Favorite spice blend (optional, to replace salt and pepper)
2 to 4 rosemary sprigs
1 tablespoon vegetable oil
1 tablespoon butter

1 Preheat the water bath to 134°F.

2 Season the lamb all over with garlic powder, salt, and pepper, or your favorite rub (if using).

3 To bag the chops, start by folding the top of a vacuum-seal or zip-top bag back over itself to form a cuff. For best results, do not bag more than four chops together in one bag. Add the chops and rosemary. Uncuff and seal using either a vacuum sealer or the displacement method.

4 Place the bag(s) in the water bath and cook for 2 to 4 hours.

5 Remove the chops from the bag(s) and place on a paper towel–lined plate. Pat dry on all sides.

6 Heat the vegetable oil in a large cast iron or stainless steel skillet over medium-high heat until oil is shimmering.

7 Add the garlic and cook for 30 seconds, until fragrant. Add the chops and rosemary to the skillet. Cook until a nice brown crust develops, about 45 seconds. Flip the chops with tongs.

8 Add the butter and melt, spooning the butter over the chops, and cook for 30 seconds. Continue until all sides are browned. Remove the chops from the heat and serve.

Tip: *If you can find goat chops, this is a simple and tasty way to prepare them!*

Korean-Style Short Ribs with Quick Kimchi

YIELD: 4 SERVINGS / ACTIVE TIME: 45 MINUTES / TOTAL TIME: 48 HOURS

These sweet, sticky braised ribs are one of my favorite dishes to cook sous vide. They are incredibly flavorful and tender and combine the amount of time you would traditionally marinate them prior to grilling with an extended low-temperature cook time, making them unbelievably succulent. With just a little effort, you can create ribs better than you can get in any Korean restaurant.

FOR THE MARINADE AND RIBS

2½ cups water
⅔ cup tamari or soy sauce
½ cup (packed) dark brown sugar
3 tablespoons apple juice
2½ tablespoons mirin (sweet Japanese rice wine) or sweet sherry
1 tablespoon sesame oil
1 pear, chopped
½ onion, chopped
1 teaspoon freshly ground black pepper
4 pounds (2 slabs, or about 8 ribs) beef short ribs
Vegetable oil
Steamed rice

FOR THE KIMCHI

1 head napa cabbage
½ cup sea salt
5 cups water
4 to 5 garlic cloves
1-inch piece ginger
1 small onion, diced
3 tablespoons fish sauce
5 tablespoons chili flakes
2 teaspoons sugar
2 tablespoons apple juice
1 teaspoon honey
2 red chiles, seeded and thinly sliced (optional)
2 tablespoons sesame seeds, toasted

TO MAKE THE MARINADE AND RIBS

1 Preheat the water bath to 144°F.

2 In a blender, add the water, tamari, sugar, apple juice, mirin, oil, pear, onion, and pepper and blend until smooth.

3 Pour the marinade into a small saucepan and heat over medium-low heat for 10 minutes.

4 To bag the ribs, start by folding the top of a vacuum-seal or zip-top bag back over itself to form a cuff. Add the ribs and marinade. Uncuff and seal the bag using either a vacuum sealer or the displacement method. If using a zip-top bag, I recommend double-bagging. If using a vacuum sealer, slowly vacuum out excess air (to prevent sucking out the marinade) and seal.

5 Place the bag in the water bath and cook for 48 hours.

6 To finish the ribs, remove the ribs from the water bath, place them on a paper towel–lined plate, and pat dry on both sides. Strain the remaining liquid through a fine-mesh strainer.

7 In a small pot, boil the liquid over high heat for 10 minutes to reduce.

8 Meanwhile, remove the bones from the ribs and trim away the fat. Cut the meat into small pieces.

9 Heat about ¼ inch of oil in a cast iron pan or deep fryer. Fry the pieces of rib meat evenly until browned, 3 to 4 minutes. Serve with the reduced sauce, kimchi, and steamed rice

TO MAKE THE KIMCHI

1 Remove the stem and core of the cabbage and trim each leaf vertically in half or in three to four strips (depending on cabbage size). Add the cabbage strips to a large mixing bowl.

2 In a small pot, combine the salt and water and bring to a full boil until the salt is dissolved. Let cool for 5 minutes and pour over the cabbage leaves. Let the cabbage sit for 10 minutes. Turn the cabbage leaves and continue to soak for 5 more minutes. Rinse once and drain thoroughly.

3 In a blender, combine the garlic, ginger, onion, and fish sauce and blend until smooth. Pour the purée into a small mixing bowl. Add the chili flakes, sugar, apple juice, and honey and mix well.

4 Add the red chiles (if using) to the cabbage, then add the sauce. Toss to coat evenly. Sprinkle in the toasted sesame seeds and toss well. Kimchi will stay fresh in the refrigerator for 10 days.

Flank Steak with Chimichurri Sauce

YIELD: 8 SERVINGS / ACTIVE TIME: 30 MINUTES / TOTAL TIME: 2 TO 24 HOURS, DEPENDING ON DESIRED TEXTURE

Flank steak isn't a cut that needs extended cooking times to break down tough connective tissues and collagen, but it definitely wants more attention than a more expensive steak like a tenderloin or strip. This recipe is a great way to turn a tougher, cheaper cut into sirloin-quality meat. Serve it on its own with chimichurri sauce, or use in tacos, sandwiches, or soups.

FOR THE MARINADE

⅓ cup olive oil
2 tablespoons red wine vinegar
⅓ cup soy sauce
¼ cup honey
2 teaspoons garlic powder
½ teaspoon freshly ground black pepper
2 pounds flank steak

FOR THE CHIMICHURRI SAUCE

½ cup red or white wine vinegar
1 teaspoon salt
4 garlic cloves, minced
1 shallot, finely chopped
¼ teaspoon red pepper flakes
½ cup minced cilantro
¼ cup minced flat-leaf parsley
2 tablespoons finely chopped fresh oregano
¾ cup extra-virgin olive oil
Salt
Freshly ground black pepper

1 Preheat the water bath to 134°F.

2 In a medium bowl, whisk together the oil, vinegar, soy sauce, honey, garlic powder, and pepper.

3 To bag the flank steak, start by folding the top of a zip-top bag back over itself to form a cuff. Add the flank steak and marinade. Uncuff and seal the bag using the displacement method.

4 Place the bag in the water bath and cook for at least 2 hours and up to 24 hours.

5 Meanwhile, make the chimichurri sauce. In a medium bowl, combine the vinegar, salt, garlic, shallot, and red pepper flakes and let stand for 10 minutes. Stir in the cilantro, parsley, and oregano, then whisk in the oil. Add salt and pepper.

6 To finish the steak, transfer the steak to a paper towel–lined plate and pat dry (but don't wipe off the marinade).

7 Heat a grill or skillet until smoking hot and add the steak. Sear quickly on each side to get a nice dark crust. Transfer to a cutting board and slice against the grain into ½-inch slices. Top with the chimichurri sauce and serve.

TRADITIONAL STYLE	134°F	2 to 3 hours
VERY TENDER	134°F	up to 24 hours

Classic Leg of Lamb with Balsamic Glaze

YIELD 8 SERVINGS / ACTIVE TIME: 25 MINUTES / TOTAL TIME: 8 TO 24 HOURS, DEPENDING ON DESIRED TEXTURE

When cooked traditionally, a large leg of lamb can come out medium-well at the outer edges and almost rare in the middle. To get a perfect medium-rare all the way through (the best way to enjoy flavorful lamb), sous vide is your only option. How chewy or tender you want it is up to you—a shorter cook time will yield a delicious, toothsome roast, and leaving it in longer will render truly succulent, tender lamb. Serve with mashed or roasted potatoes (page 154 or 157) for a feast that will impress all your guests.

1 (5- to 7- pound) boneless leg of lamb
½ to 1 cup Lamb Rub (page 219)
1 tablespoon olive oil
2 cups balsamic vinegar

1 Preheat the water bath to 131°F for traditional style or 134°F for very tender.

2 If your lamb has a thick layer of fat, score through the fat with a sharp knife in parallel lines. Repeat in the opposite direction to create a diamond pattern. This releases the connective tissue tension on the surface and allows your rub to penetrate more deeply into the flesh.

3 Rub the leg all over with the spice rub. To bag the lamb, start by folding the top of a vacuum-seal or zip-top bag back over itself to form a cuff. Add the lamb and oil. Uncuff and seal the bag using a vacuum sealer or the displacement method. If using a zip-top bag, I recommend double-bagging.

4 Place the bag in the water bath and cook for at least 8 hours and up to 24 hours. The longer time will result in more tender meat.

5 Remove the lamb from the bag and reserve the juices.

6 Preheat the oven to 450°F.

7 In a saucepan, cook the reserved juices and balsamic vinegar over medium-high heat until the mixture reduces by half, about 20 minutes. The undercooked proteins are going to coagulate and thicken, so don't be surprised when you see it.

8 Strain the glaze through a fine-mesh sieve lined with damp paper towels.

9 Place a rack on a cooking tray lined with foil. Set the lamb on the rack and transfer to the oven to crisp up the skin, 10 to 15 minutes. Drizzle with the balsamic glaze and serve.

Lamb Shoulder Carnitas Tacos

YIELD: 4 TO 6 SERVINGS / ACTIVE TIME: 30 MINUTES / TOTAL TIME: 72 HOURS, 20 MINUTES

This recipe comes to us from Paul Palop, sous vide enthusiast and blogger at That Other Cooking Blog, where he features a wide array of sous vide recipes. I'll let Paul describe the recipe: "These lamb shoulder tacos pair well with hoisin sauce and adobo sauce, the core of the popular chipotle sauce. It's really spicy but has an incredible smokiness. Mix hoisin and adobo together and you'll have a pretty sexy combination. Just call these 'Mexi-Asian' lamb tacos." And the mayonnaise? "I just love mayo, and it goes really well with the lamb, but you can leave it out if you like."

FOR THE LAMB

1 pound bone-in lamb shoulder chops
2 to 3 tablespoons kosher salt
1 tablespoon vegetable oil, plus more for frying
1½ teaspoons hoisin sauce
1½ teaspoons chipotles in adobo sauce, chopped fine

FOR THE PICO DE GALLO

2 large or 3 small tomatoes, seeded and diced (about 1½ cups)
¼ cup diced red onion
1 teaspoon diced jalapeño
1 garlic clove, minced
Juice of 2 limes
2 tablespoons minced cilantro
Salt and pepper to taste

FOR THE PICKLED ONIONS

⅓ cup white wine vinegar
¼ cup water
1 tablespoon sugar
1 teaspoon salt
1 cup diced red onion

FOR THE ASSEMBLY

12 small flour tortillas
Mayonnaise (optional)
½ cup cilantro leaves, for garnish

TO MAKE THE LAMB

1 Dust the lamb shoulder chops generously with salt. Allow them to cure overnight in an airtight zip-top bag in the refrigerator.

2 The next day, preheat the water bath to 134°F.

3 Rinse the lamb and the bag. Return the lamb to the zip-top bag and add 1 tablespoon of vegetable oil. Seal the bag using the displacement method.

4 Cook for 72 hours.

5 After the lamb is cooked, remove it from the bag. Discard the juices or reserve for another purpose.

6 Dry the lamb pieces with a paper towel.

7 Heat about 2 inches of vegetable oil in a heavy pot to 375°F. Add the lamb chops and fry for 3 minutes, or until crisp.

8 Remove from the pot and drain on a rack placed over a sheet pan or paper towels for a few minutes.

9 Shred the lamb into a large bowl. Add the hoisin and chipotles and mix well.

TO MAKE THE PICO DE GALLO

1 In a small bowl, mix the tomatoes, onion, jalapeño, garlic, lime juice, cilantro, salt, and pepper together.

2 Allow to rest for 20 minutes.

TO MAKE THE PICKLED ONIONS

1 Add the vinegar, water, sugar, and salt to a small pot and bring to a simmer over medium-low heat.

2 Add the onion. Simmer for a couple of minutes.

3 Remove from the stove. Cool for a few minutes, then refrigerate until chilled.

TO ASSEMBLE

1 Heat a small skillet over medium-high heat. Place a tortilla in the skillet and cook until it starts to puff and brown, about 30 seconds. Flip and cook the other side. Repeat with the remaining tortillas.

2 Spread each tortilla with a teaspoon or so of mayonnaise (if using). Add about 2 tablespoons of lamb, then top with the pico de gallo and and pickled onions.

3 Garnish with some cilantro leaves.

Ragu Bolognese

YIELD: 6 TO 8 SERVINGS / ACTIVE TIME: 40 MINUTES / TOTAL TIME: 5 HOURS, 40 MINUTES

This recipe gives the same depth and layers of flavors as the traditional version, including the smoky flavors from pre-seared meat and vegetables and herbs that maintain their freshness in both texture and taste. I used to make this when I had hours of uninterrupted time. But with this method, I do the first steps before I go to bed, pop it in the refrigerator, and put it in to cook in the morning. When I get home, dinner is ready.

4 tablespoons olive oil, divided
6 ounces chopped pancetta or bacon
1 large onion, finely chopped
1 carrot, finely chopped
2 celery stalks, finely chopped
2 large garlic cloves, minced
1 pound ground beef
1 pound ground pork
1 (14-ounce) can Italian chopped tomatoes
1 (6-ounce) can tomato paste
¾ cup red wine
1 bay leaf
4 tablespoons olive oil
4 tablespoons chopped basil, divided
Salt
Freshly ground black pepper
Dash freshly grated nutmeg

1 Heat 3 tablespoons of oil in a Dutch oven over medium heat. Add the pancetta. Cook until it starts to render its fat, about 1 minute.

2 Add the onion, carrot, celery, and garlic and continue cooking, stirring occasionally, until completely softened and light golden brown, about 10 minutes. Set aside on a separate plate.

3 Wipe out the pan and heat ½ tablespoon of oil over high heat until it shimmers. Add the beef and brown for 3 to 4 minutes, stirring occasionally. Transfer to a plate.

4 Heat the remaining ½ tablespoon of oil and brown the ground pork for 3 to 4 minutes, stirring occasionally.

5 Return the beef and the onion mixture to the pan with the pork. Add the tomatoes, tomato paste, red wine, bay leaf, 2 tablespoons of the basil, salt, pepper, and nutmeg, and stir well.

6 Using a ladle, carefully transfer the mixture to a zip-top bag, and seal using the displacement method.

7 Place the bag in the water bath and cook for 5 hours.

8 When it's ready, serve the sauce with your favorite fresh pasta. If you're making a double batch, chill half in an ice bath before storing in the freezer for up to 1 month.

Tip: *Browning the meat in batches ensures that the temperature of the oil doesn't drop and that there is enough surface area for the meat to brown evenly.*

Tender Meatballs

YIELD: 4 SERVINGS / ACTIVE TIME: 10 MINUTES / TOTAL TIME: 1½ TO 2 HOURS

The amazing thing about cooking meatballs this way is that you can make them as large as you want and not have to worry about whether or not they're cooked all the way through. Imagine baseball-size meatballs! That dream can be yours when you cook sous vide. Combine beef, pork, and lamb for a truly flavorful dish, or stick to basic beef. These are great with a tomato sauce, gravy, or sweet cocktail sauce.

1 pound mixed ground beef, pork, and/or lamb
½ cup milk
½ cup bread crumbs
1 large egg
½ cup grated Parmesan cheese
2 tablespoons chopped fresh parsley
1 teaspoon garlic powder
½ teaspoon dried oregano
1 teaspoon salt
½ teaspoon freshly ground black pepper
1 tablespoon oil

Tip: *Add meatballs to your favorite pasta dish, enjoy them in a sandwich, or drop them in soup. Any way you use them, you can't go wrong.*

1 Preheat the water bath to 140°F.

2 In a large bowl, mix together the ground meats, milk, bread crumbs, egg, Parmesan cheese, parsley, garlic, oregano, salt, and pepper until all are evenly distributed.

3 Shape the mixture into 16 meatballs about 1 inch in diameter (or 4 baseball-size meatballs). Place 8 meatballs (or 2 large meatballs), each in a single layer, in two zip-top bags and seal using the displacement method.

4 Place the bags in the water bath and cook for 1½ hours (baseball-size meatballs will take up to 2 hours).

5 When the meatballs are done cooking, remove from the bags.

6 Heat the oil in a large nonstick skillet over medium-high heat. When the oil is shimmering, add the meatballs and quickly sear on all sides. Serve.

Chili

YIELD: 8 SERVINGS / ACTIVE TIME: 15 MINUTES / TOTAL TIME: 2 HOURS

I grew up where chili meant chili beans, tomatoes, and chili powder, so if you're looking for classic Texas chili con carne, this isn't for you. But this recipe is highly customizable to match whatever secret recipe you've had handed down to you from Great-Uncle Jim. Use this as a base to dress up some leftover flank (the added cooking time won't hurt) or start fresh with ground beef like we do at home.

1 cooked Flank Steak (page 60), chopped, or 2 pounds uncooked ground chuck
3 tablespoons oil or fat (if using uncooked ground chuck)
1 small onion, diced
1 garlic clove, minced
2 tablespoons chili powder
1 tablespoon smoked paprika
1 tablespoon ground cumin
1 teaspoon garlic powder
1 teaspoon onion powder
1 cup thinly sliced scallion, green bottoms and white tops separated
1 (14.5-ounce) can kidney beans, drained
1 (14-ounce) can fire-roasted tomatoes
1 (4-ounce) can fire-roasted chiles
1 cup Beef Stock (page 206)
Salt
Freshly ground black pepper
1 cup shredded Monterey Jack cheese, for garnish
1 cup sour cream, for garnish
1 bunch cilantro, chopped, for garnish

1. Preheat the water bath to 140°F.
2. If using cooked flank steak, skip to step 3. If using ground chuck, heat the oil in a stainless steel or cast iron skillet until almost smoking. Add the ground beef and cook until brown, 2–5 minutes.
3. In a large bowl, mix together the onion, garlic, chili powder, paprika, cumin, garlic powder, onion powder, beans, tomatoes, chiles, white scallion bottoms, and stock. Add the ground beef or flank steak and stir gently to combine.
4. Evenly distribute the chili into two large zip-top bags and seal using the displacement method.
5. Place the bags in the water bath and cook for 2 hours.
6. Remove from the water bath and season with salt and pepper.
7. Serve the chili in bowls topped with some of the Monterey Jack cheese, sour cream, cilantro, and green scallion tops.

Tip: *This recipe is even more delicious with ground buffalo meat if you can find it in your grocer's freezer!*

Lamb Curry

YIELD: 6 SERVINGS / ACTIVE TIME: 20 MINUTES / TOTAL TIME: 16 TO 24 HOURS

Turn up the heat with this curry inspired by the Kashmiri lamb dish *rogan josh*. Add more cayenne or toss in some fresh chilies for maximum strength, or dial it down for more delicate palates. The high temperature here is ideal for breaking down tough cuts like lamb shoulder or leg into a tender stew, but if you prefer to cook with a loin, turn the temperature down to 140°F. I recommend pre-searing the meat to let the great flavors penetrate the sauce while cooking.

2 pounds boneless lamb shoulder
4 tablespoons canola oil, divided
2 onions, thinly sliced (3 cups)
2 garlic cloves, minced
2 tablespoons minced fresh ginger
1 tablespoon curry powder
1 teaspoon ground turmeric
½ teaspoon cayenne pepper
2 bay leaves
1 (14-ounce) can tomato purée
1 teaspoon garam masala
Salt
Freshly ground black pepper
Cilantro leaves, for garnish
Cooked basmati rice and warm naan, for serving

1 Preheat the water bath to 180°F.

2 Pat the lamb dry with paper towels.

3 Heat 2 tablespoons of oil in a large cast iron or stainless steel skillet over high heat until the oil is almost smoking. Add the lamb and turn to brown on all sides, just to get a crust on it, 3 to 5 minutes each side. Transfer to a plate.

4 Heat the remaining 2 tablespoons of oil in the skillet over medium-high heat until shimmering.

5 Add the onions and cook until just translucent, 3 to 5 minutes.

6 Add the garlic, ginger, curry powder, turmeric, cayenne, and bay leaves and cook for 30 seconds.

7 To bag the lamb, start by folding the top of a large zip-top bag back over itself to form a cuff. Add the seared lamb, the onion-spice mixture, and the tomato purée. Uncuff and seal using the displacement method.

8 Place the bag in the water bath and cover the water bath with plastic wrap to minimize water evaporation. Add water intermittently to keep the bag submerged.

9 Cook for 16 to 24 hours. Pour the contents of the bag into a large sauce pan or Dutch oven. Remove the lamb from the stew. Discard the bay leaves and any large pieces of fat. Shred the lamb and return to the stew.

10 Stir in the garam masala, and season with salt and pepper. Reheat if necessary over low heat.

11 Garnish with the cilantro leaves and serve with the rice and naan.

Tip: *This is a wonderful dish to try with goat instead of lamb.*

Pho with Flank Steak and Bone Broth

YIELD: 4 SERVINGS / ACTIVE TIME: 45 MINUTES / TOTAL TIME: 45 MINUTES (PLUS UP TO 26 HOURS FOR MAKING THE BROTH)

Vietnamese pho (pronounced *fuh*) is a hearty, meaty broth with noodles, beef, and lots of fresh herbs. It's just the kind of thing you crave on a chilly day or when you feel a cold coming on. This dish is a great way to use up leftover flank steak and showcase delicious bone broth. Everyone eats it their own way—mix everything in all at once or top each spoonful with herbs and hot sauce individually.

- 6 cups Bone Broth (page 206)
- 2 large onions, quartered
- 4-inch piece ginger, peeled and cut in half lengthwise
- 2 (3-inch) whole cinnamon sticks
- 2 whole star anise
- 3 whole cloves
- 2 teaspoons coriander seeds
- 1 tablespoon soy sauce (substitute tamari if making gluten-free)
- 1 tablespoon fish sauce
- Flank Steak (page 60), thinly sliced
- 1 pound rice noodles (dried or fresh)
- 1 cup fresh mung bean sprouts, plus more for topping
- 1 small bunch mint, roughly chopped
- 1 small bunch cilantro, roughly chopped
- 1 small bunch basil, roughly chopped
- 1 lime, cut into wedges
- 1 jalapeño, thinly sliced
- 2 scallions, green and white parts thinly sliced, for topping
- Hoisin sauce, for topping
- Sriracha hot sauce, for topping

1 Preheat the broiler.

2 In a medium saucepan over low heat, bring the broth to a simmer.

3 Put the onions and ginger on a baking sheet and char under the broiler for 5 to 10 minutes.

4 Add to the broth the onions-ginger mixture, cinnamon, anise, cloves, coriander, soy sauce, and fish sauce.

5 Cover and continue to simmer for 30 minutes, allowing the aromatics to infuse into the broth.

6 While the broth is simmering, put the beef on a plate, cover with plastic wrap, and freeze for 15 minutes until firm to the touch. This will make it easier to slice the beef thinly.

7 Meanwhile, cook the rice noodles according to package instructions. Strain the noodles and run them under cool water to stop them from cooking. The noodles will start to stick together after cooking, so divide them immediately among 4 serving bowls.

8 Remove the beef from the freezer and, using your sharpest knife, slice it into ¼-inch slices against the grain.

9 Place the bean sprouts, mint, cilantro, basil, lime, and jalapeño in separate mounds on a serving platter.

10 When the broth is ready, strain and discard the solids from the broth.

11 Arrange the beef on top of the noodles in the bowls, ladle the broth into each bowl, and serve with the toppings alongside so each guest can customize to their taste.

Beef Tenderloin with Red Wine Reduction

YIELD: 6 TO 8 SERVINGS / ACTIVE TIME: 30 MINUTES / TOTAL TIME: 1½ TO 2 HOURS

When preparing an expensive cut of meat like a beef tenderloin, you want to do everything you can to ensure you're serving a perfectly prepared meal. Because this cut is so lean, it's easy to undercook or toughen it by using traditional cooking methods. So whether you're slicing this into steaks or presenting it as a pastry-encrusted Wellington, sous vide is your key to a flawless filet.

BEEF TENDERLOIN

2 pounds center-cut beef tenderloin
2 tablespoons vegetable oil, for searing
1 tablespoon butter
4 thyme sprigs
2 rosemary sprigs

RED WINE REDUCTION

2 tablespoons olive oil
4 large shallots, sliced
12 whole black peppercorns
1 bay leaf
1 thyme sprig
Splash red wine vinegar
1 bottle dry red wine (I like Shiraz or Zinfandel)
3 cups Beef Stock (page 206)
Salt
Freshly ground black pepper
1 tablespoon butter, for searing

1 Preheat the water bath to 138°F.

2 Truss the tenderloin with butcher's twine in three intervals to help it keep its shape.

3 Heat the oil in a stainless steel or cast iron skillet over high heat until the oil begins to shimmer. Add the loin and sear on all sides until the outside is brown and crisp, 2 to 3 minutes per side.

4 Remove the loin from the pan and set aside.

5 Add the butter to the pan and sauté the herbs until bright green, 30 seconds to 1 minute.

6 To bag the tenderloin, start by folding the top of a vacuum-seal or zip-top bag back over itself to form a cuff. Add the tenderloin and the herbs and cooking liquids. Uncuff and seal the bag using a vacuum sealer or the displacement method.

7 Place the bag in the water bath and cook for 90 minutes to 2 hours.

8 Meanwhile, make the wine sauce. Heat the oil in a stainless steel or cast iron pan. Add the shallots, peppercorns, bay leaf, and thyme and continue to cook, stirring frequently, until the shallots turn golden brown, about 5 minutes.

9 Add the vinegar and let simmer for a few minutes until almost dry. Add the wine and boil until reduced by half, about 20 minutes. Add the stock and bring to a boil again. Lower the heat and simmer gently for 1 hour, occasionally skimming the surface of the sauce. Strain the liquid through a fine sieve. Season with salt and pepper and set aside.

10 Remove the tenderloin from the bag, reserving the cooking liquids and herbs. Heat the butter in a stainless steel or cast iron skillet over medium-high heat and add the cooking liquid and herbs from the cooking bag. When the bubbling slows down, add the loin to the skillet for a final sear, quickly crisping the outside for about 30 seconds on all sides while spooning the hot liquid over the top.

11 Cut into thick slices, drizzle with the wine sauce, and serve.

Beef Wellington

YIELD: 8 SERVINGS / ACTIVE TIME: 20 MINUTES / TOTAL TIME: 1½ HOURS

Beef Wellington is one of the most elegant, impressive, and complex dishes you can serve. Fortunately, sous vide takes all of the guesswork out of the preparing the tenderloin, so you can focus on getting that perfect golden finish to your pastry.

1 tablespoon olive oil
1 cup wild mushrooms, finely chopped
Thyme leaves from 1 sprig
Sea salt
Freshly ground black pepper
1 sheet frozen puff pastry, thawed
1 cooked Beef Tenderloin (page 74)
4 slices prosciutto
1 egg yolk, beaten with 2 teaspoons water and a pinch salt
Red Wine Reduction (from Beef Tenderloin recipe, page 74)

1 Heat the oil in a stainless steel or cast iron skillet over medium heat. Add the mushrooms and cook, stirring, for 2 minutes. Season with thyme, salt, and pepper and cook until all the excess moisture has evaporated and the mushrooms form a paste, about 10 minutes. Remove the paste from the pan and allow to cool.

2 Place the puff pastry on a lightly floured surface and roll into a rectangle large enough to envelop the entire tenderloin. Chill the pastry rectangle in the refrigerator.

3 Lay a large sheet of plastic wrap on a work surface and place 4 slices of prosciutto in the middle, overlapping slightly along the long edges. Spread half the mushroom paste evenly on top.

4 Place the tenderloin on top of the mushroom-covered prosciutto. Using the plastic wrap, roll the prosciutto over the beef, then roll and wrap the plastic film around the beef to get a nice, evenly thick log. Chill for 30 minutes.

5 Remove the pastry from the refrigerator and brush with the egg wash. Remove the beef from the refrigerator and remove the plastic film from the beef. Wrap the pastry around the entire tenderloin, sealing the seam and the ends.

6 Trim away any excess pastry and brush the packet all over with the egg wash.

7 Cover with plastic wrap and chill for about 30 minutes. Meanwhile, preheat the oven to 425°F.

8 Remove the packet from the refrigerator and remove the plastic wrap. Score the pastry lightly 4 or 5 times with a sharp knife and brush with more egg wash.

9 Bake for 10 to 15 minutes, until the pastry is golden brown. Rest for 10 minutes before carving.

10 Slice and serve with the red wine reduction on the side.

4

POULTRY AND PORK

As with eggs, the different parts of poultry reach their ideal doneness at different temperatures. When cooking a whole chicken the traditional way, this means you're setting your goal for the dark meat to be cooked through, which turns the light meat tough and stringy. Even when cooking chicken breasts on their own, it's impossible to create the delicate texture you can achieve with sous vide. By cooking at lower temperatures, you're preventing water loss, resulting in amazingly moist meat. In the case of pork, the various cuts should be treated differently. Tender cuts cook more quickly and at lower temperatures than larger roasts. Pork tenderloin comes out tender and buttery, while a pork butt or shoulder can be almost magically converted to meat just as succulent as a more expensive chop.

Basic Chicken Breasts

YIELD: 1 TO 6 CHICKEN BREASTS / ACTIVE TIME: 10 MINUTES / TOTAL TIME: 1 TO 3 HOURS

Chicken breasts cooked to 140°F have a very tender, extremely juicy, and smooth texture that is firm without any stringiness or tackiness. Many people used to traditionally cooked chicken can find the different texture off-putting, so I recommend a slightly higher temperature. If preparing chicken to be served cold or in a salad, however, I find 140°F to be ideal. Experiment to find your perfect doneness. I prefer skin-on, bone-in chicken breasts, but boneless and skinless will work just as well.

1 to 6 skin-on, bone-in chicken breasts
Salt (optional)
Freshly ground black pepper (optional)
1 thyme or oregano sprig per breast (optional)
2 to 3 lemon slices per breast (optional)
1 tablespoon olive oil

1. Preheat the water bath to 146°F.
2. Season the chicken breasts with salt and pepper (if using).
3. To bag the chicken breasts, start by folding the top of a vacuum-seal or zip-top bag back over itself to form a cuff. This prevents chicken juices from getting on the edges of the bag, which can interfere with the seal or spread bacteria. For best results, do not bag more than two pieces together in one bag.
4. Add the chicken breasts and any aromatics, such as fresh herbs or lemon slices (if using). Uncuff and seal the bag(s) using either a vacuum sealer or the displacement method.
5. Place the bag(s) in the water bath and cook for 1 to 3 hours. After 3 hours, the texture of the chicken breast can become chalky.

6 After cooking, remove the chicken breasts from the bag, discard any aromatics, and place the chicken on a paper towel–lined plate. Pat dry.

7 Heat the oil in a large cast iron or stainless steel skillet over medium-high heat until the oil is slightly shimmering. Add the chicken to the hot skillet, skin-side down (if using skin-on chicken breasts). Cook for about 1 minute, swirling the oil or pressing down on the chicken to ensure the most contact with the hot oil for maximum browning. Flip the chicken and cook for another 1 minute. Remove from the heat.

8 If using bone-in chicken breasts, remove the bones before serving. You should be able to peel the breast off the breastbone by simply running your thumb in between the meat and the bone.

9 Cut the chicken into thick slices on the bias, keeping the skin intact, and serve.

Tip: *Buy chicken breasts in bulk and bag them up separately with different seasonings. Cook them all at once, then chill or use them in quick lunches and dinners you won't get tired of.*

Basic Chicken Thighs

YIELD: 1 TO 6 CHICKEN THIGHS / ACTIVE TIME: 15 MINUTES / TOTAL TIME: 2 TO 6 HOURS

Chicken thighs require high cooking temperatures to break down their connective tissues and create juicy, tender meat. I suggest a 2- to 4-hour cooking time. After 4 hours, the meat becomes less juicy but is still tender up until the 6-hour mark. Unlike breasts, thighs have a layer of fat that makes it hard to get a great, crispy skin. Either use skinless thighs, or follow the steps below to remove the gelled fat before searing.

1 to 6 skin-on, bone-in chicken thighs
Salt (optional)
Freshly ground black pepper (optional)
1 sprig each thyme and oregano per thigh (optional)
2 to 3 lemon slices per thigh (optional)
1 tablespoon olive oil

1 Preheat the water bath to 165°F.

2 Season the chicken thighs with salt and pepper (if using).

3 To bag the chicken thighs, start by folding the top of a vacuum-seal or zip-top bag back over itself to form a cuff. This prevents chicken juices from getting on the edges of the bag, which can interfere with the seal or spread bacteria. For best results, do not bag more than two pieces together in one bag.

4 Add the chicken thighs and any aromatics, such as fresh herbs or lemon slices (if using). Uncuff and seal the bag(s) using either a vacuum sealer or the displacement method.

5 Place the bag(s) in the water bath and cook for 2 to 6 hours.

6 Once the chicken has cooked for the desired time, prepare an ice bath. Transfer the bagged chicken to the ice bath to chill completely. At this stage, the chicken can be stored in the refrigerator for up to 3 days before finishing and serving.

7 Remove the chicken from the bag, discard any aromatics, and place on a paper towel–lined plate.

8 Remove any gelled juices from the chicken with your fingers and discard.

9 Heat the oil in a large cast iron or stainless steel skillet over medium-high heat until the oil is slightly shimmering. Add the chicken to the hot skillet, skin-side down (if using skin-on chicken thighs), and cook for 4 to 5 minutes, pressing down on the chicken to ensure the most contact with the hot oil for maximum browning. Flip the chicken and cook for another 1 minute. Remove from the heat and serve.

Tip: *Reuse the water multiple times, but do remove the circulator and cover the water bath between uses. Do this only for a few days and as long as the water remains sterile. Dirty water can be used to water plants!*

Crispy Chicken Thighs with Mustard-Wine Sauce

YIELD: 4 SERVINGS / ACTIVE TIME: 20 MINUTES / TOTAL TIME: 2 TO 8 HOURS

This chicken thigh recipe gives you all the flavor you could ever want in your poultry, topped off with a mustard-wine sauce made with lemon, butter, and fresh herbs (if that's your jam). Not only will your chicken be flavorful, it will have pristine texture and consistency: deliciously crispy on the outside and juicy and tender on the inside.

4 skin-on, bone-in chicken thighs
Kosher salt
Freshly ground black pepper
4 thyme or rosemary sprigs (optional)
1 tablespoon canola, vegetable, or grapeseed oil
1 small shallot, minced
1 cup dry white wine
1 tablespoon Dijon mustard
2 tablespoons butter
½ teaspoon freshly squeezed lemon juice
2 tablespoons minced fresh parsley, divided

1 Preheat the water bath to 165°F.

2 Generously season the chicken with salt and pepper.

3 To bag the chicken thighs, start by folding the top of a vacuum-seal or zip-top bag back over itself to form a cuff. This prevents chicken juices from getting on the edges of the bag, which can interfere with the seal or spread bacteria. For best results, do not bag more than two pieces together in one bag.

4 Add the chicken thighs and thyme or rosemary sprigs (if using). Uncuff and seal the bag(s) using either a vacuum sealer or the displacement method.

5 Place the bag(s) in the water bath and cook the chicken for 2 to 4 hours for juicy chicken and up to 8 hours for less juicy but fall-off-the-bone tender meat.

6 When the chicken is cooked, transfer the bags to an ice bath and chill completely.

7 Proceed to the next step or store the chicken in the refrigerator for up to 4 days before finishing.

8 Remove the chicken from the bags, reserving any gelled liquid, and pat thoroughly dry with paper towels.

9 Heat the oil in a nonstick or cast iron skillet over medium heat until shimmering. Gently lay the chicken in the skillet, skin-side down, using your fingers or a set of tongs to press flat. Cook until brown and crisp, about 8 minutes, reducing the heat if necessary. Flip the chicken and cook for about 2 minutes more. Transfer the chicken to a paper towel–lined plate.

10 Increase the heat to medium-high. Add the shallot and cook, stirring, until fragrant, about 30 seconds.

11 Add the wine and cook until reduced by half, about 2 minutes. Stir in the gelled chicken juices and mustard. Remove from the heat and whisk in the butter, lemon juice, and 1 tablespoon of parsley. Season with salt and pepper.

12 Serve the chicken immediately with pan sauce and top with the remaining 1 tablespoon of parsley.

Tip: *This dish is great with the Perfect Roasted Potatoes (page 157)!*

Juiciest Fried Chicken

YIELD: 4 SERVINGS / ACTIVE TIME: 30 MINUTES / TOTAL TIME: 1 TO 5 HOURS

Should you brine your chicken before cooking sous vide? The idea is that brining the chicken makes it more succulent and flavorful. I tried it (using Thomas Keller's brine solution, no less) and came to the conclusion that sous vide's delivery of succulent, juicy chicken made brining unnecessary even when going for the juiciest fried chicken on the planet. In any event, I include a brine solution here for you to try in case you'd like to compare. When it comes to frying, the secret is the crumb. Here, I've added a little buttermilk to the flour to form a really nice, large crumb. I also add baking powder, because as it fries, the baking powder releases carbon dioxide, leavening the crust and increasing its surface area, keeping it light and crisp.

FOR THE BRINE (OPTIONAL)

- 8 cups water
- ⅓ cup kosher salt
- 3 tablespoons honey
- 6 bay leaves
- 10 unpeeled garlic cloves
- 1 tablespoon whole black peppercorns
- 2 large rosemary sprigs
- ½ bunch thyme
- ½ bunch parsley
- 3 teaspoons finely grated lemon peel
- ¼ cup lemon juice

Tip: *Don't drain your chicken on paper towels! The paper towels will just cause it to steam and get soggy. Get out a wire rack and set it over the towels or a baking sheet so the chicken can stay crispy while it releases any extra oils.*

FOR THE CHICKEN

- 1 whole chicken, cut into pieces (or a mix of your favorite parts)
- 1 teaspoon baking powder
- 1 tablespoon garlic powder
- 1 tablespoon onion powder
- 1 teaspoon smoked paprika
- 1 teaspoon cayenne pepper
- 1 teaspoon kosher salt
- ½ teaspoon freshly ground black pepper
- 1½ cups all-purpose flour
- ½ cup cornstarch
- 1 cup buttermilk, divided
- 12 cups peanut oil for deep frying (less for pan frying)

1 If you're brining, mix all the ingredients together in a pan and bring to a boil. Stir to dissolve the salt. Transfer to the refrigerator and cool completely. (Don't be tempted to add the chicken to slightly warm water. This is a safety no-no. Be sure the brine is completely room temperature.) Transfer the brine to a large bowl and submerge the chicken. Refrigerate overnight.

2 Preheat the water bath to the desired final temperature.

3 To bag the chicken, start by folding the top of a vacuum-seal or zip-top bag back over itself to form a cuff. This prevents chicken juices from getting on the edges of the bag, which can interfere with the seal or spread bacteria. For best results, do not bag more than two pieces together in one bag.

4 Remove the chicken pieces from the brine (if using) and pat dry. Discard the brine. Add the chicken to the bags. Uncuff and seal the bags using either a vacuum sealer or the displacement method.

5 Place the bags in the water and cook following the chart below.

6 Remove the chicken from the bags and pat thoroughly dry with paper towels.

7 To make the spice mixture, combine the garlic powder, onion powder, smoked paprika, cayenne, and pepper together in a small bowl.

8 Toss the flour, cornstarch, baking powder, and 2 tablespoons of the spice mixture together in a zip-top bag. Remove half the flour mixture and spread it out on a large plate.

9 In a small bowl, whisk together the buttermilk, salt, and remaining spice mixture.

10 To the remaining flour mixture the bag, add just enough buttermilk, about 1 tablespoon at a time, to create a good-size crumb (about the size of crumbled feta cheese).

11 To fry, heat the oil to 375°F in a deep fryer, or to pan fry, heat 2 inches of oil in a cast iron skillet or high-sided pan. Place a wire rack over a baking sheet.

12 Dip the chicken pieces in the flour mixture on the plate, turning to coat. Repeat with the buttermilk mixture in the bowl. Finally, toss the chicken in the bag with the flour-buttermilk crumb. Repeat until all pieces are breaded.

13 Fry the chicken in batches in hot oil for 3 minutes on each side (don't crowd the oil). Remove from the oil and place on the prepared wire rack to cool.

LIGHT AND DARK MEAT	148°F	3 to 5 hours
LIGHT MEAT ONLY	144°F	1 to 2 hours
DARK MEAT ONLY	167°F	1½ to 3 hours

Fried Chicken Sliders with Sriracha Mayo

YIELD: 4 SERVINGS (8 SLIDERS) / ACTIVE TIME: 30 MINUTES / TOTAL TIME: 3½ HOURS

Fried chicken is one of life's greatest pleasures. Gourmet comfort food is a very real trend, but some home chefs question their ability to pull it off. With this sous vide recipe, you can rest assured that your chicken thighs will come out tender, juicy, and bursting with flavor. Topped with sriracha mayo and dill pickles, these little sliders are as gastronomically impressive as they are casually fun to eat.

4 boneless, skinless chicken thighs
1 cup dill pickle juice
¼ cup mayonnaise
1 tablespoon sriracha, plus additional for dredging if desired
Frying oil
1 cup flour
1 tablespoon paprika
1 tablespoon onion powder
1 teaspoon garlic powder
1 teaspoon salt
1 teaspoon freshly ground black pepper
1 egg
1 cup milk
8 potato rolls
8 to 16 dill pickle slices

1 Lightly pound the chicken thighs to achieve an even thickness.

2 Preheat the water bath to 167°F.

3 To bag the chicken thighs, start by folding the tops of two vacuum-seal or zip-top bags back over themselves to form cuffs. This prevents chicken juices from getting on the edges of the bag, which can interfere with the seal or spread bacteria.

4 Add two chicken thighs and half the pickle juice to each bag. Uncuff and seal using the vacuum sealer on the gentle setting or the displacement method.

5 Place the bags in the water bath and cook the chicken for 3 hours.

6 Combine the mayonnaise and 1 tablespoon of sriracha.

7 When the chicken is cooked, remove it from the bags, place it on a paper towel–lined plate, and pat dry. Cut each thigh in half.

8 In a medium bowl, mix together the flour, paprika, onion powder, garlic powder, salt, and pepper. Transfer half the flour mixture to another bowl. Whisk together the egg and milk in a third bowl, and add 1 to 2 teaspoons of sriracha (if using). To one of the bowls of the flour mixture, add just enough of the milk mixture (about 1 tablespoon) to make a coarse crumb (about the size of crumbled feta cheese).

9 Dredge each piece of chicken through the flour mix, then the milk mixture, then the crumb coating.

10 Heat about 1 inch of oil in a cast iron skillet to about 350°F. Working in batches, fry the chicken pieces in the oil until they've got a gorgeous golden crust, 2 to 3 minutes. (Remember, the inside is already cooked just the way you want it, so as soon as the crust looks good, remove them from the oil.)

11 Cut the rolls in half and toast.

12 Spread the sriracha mayo on the top halves of the rolls.

13 Place a piece of chicken on each bottom half, and top with 2 to 4 dill pickle slices and the bun top. Serve two sliders to a plate.

Tip: *For a truly authentic fast-food imitation, look for Kewpie mayo!*

Korean Chicken Wings with Barbecue Sauce

YIELD: 4 TO 6 SERVINGS / ACTIVE TIME: 20 MINUTES / TOTAL TIME: 2½ HOURS

Hot wings are a crowd pleaser, but this recipe takes them to a new level with a Korean-inspired barbecue sauce. The result is tangy, sweet, spicy wings that basically melt off the bone. You'll never look at chicken wings the same way again.

3 pounds chicken wings
Salt
Freshly ground black pepper

FOR THE SAUCE

¼ cup soy sauce
¼ cup Korean chili sauce or sriracha
2 tablespoons rice vinegar
2 tablespoons honey
1 tablespoon sesame oil
2 teaspoons garlic powder
1 teaspoon ground ginger
Freshly ground black pepper

TO SERVE

1 scallion, green and white parts thinly sliced
1 tablespoon chopped peanuts

1 Preheat the water bath to 145°F.

2 Season the chicken wings with salt and pepper. To bag them, start by folding the top of a vacuum-seal or zip-top bag back over itself to form a cuff. This prevents chicken juices from getting on the edges of the bag, which can interfere with the seal or spread bacteria.

3 Add the chicken. Uncuff and seal the bag using the displacement method or a vacuum sealer on the moist setting.

4 Place the bag in the water bath and cook for 2 hours.

5 Meanwhile, make the barbecue sauce. In a medium saucepan over medium heat, whisk together the soy sauce, chili sauce, vinegar, honey, sesame oil, garlic, ground ginger, and a dash pepper. Simmer until the mixture is reduced by half. Transfer to a large bowl and set aside.

6 When the chicken is done, remove the bag from the water bath. Remove the wings from the bag and pat dry. Discard the cooking liquid. Heat the broiler to high.

7 Transfer the wings to the bowl with the sauce and toss to coat. Spread the wings and any remaining sauce onto a foil-lined rimmed baking sheet.

8 Broil the wings until they are slightly charred and the sauce is sticky, 5 to 10 minutes.

9 Serve the wings topped with the scallions and peanuts.

Tip: *To make classic hot wings, cook as instructed, replacing the Korean-style sauce with the Classic Hot Wing Sauce recipe on page 213.*

Chicken Tikka Masala

YIELD: 4 SERVINGS / ACTIVE TIME: 35 MINUTES / TOTAL TIME: 1½ TO 2½ HOURS

Though it originated in the United Kingdom, this Chicken Tikka Masala is rich with traditional Indian flavors and so simple to put together. It's a fan favorite, requested by Indian cuisine lovers in restaurants around the world. This flavorful dish has international appeal and plenty of flavor to match. You can adjust the seasonings to make it as spicy or as mild as you want, and serve it with naan and rice for an easy, filling meal.

1½ tablespoons ground cumin
1½ tablespoons paprika
1 tablespoon ground coriander seed
1 teaspoon garlic powder
1 teaspoon ground ginger
1 teaspoon ground turmeric
½ teaspoon cayenne pepper
¼ teaspoon kosher salt
4 boneless, skinless chicken breasts (about 2 pounds)
5 tablespoons butter or ghee, divided
1 small onion, thinly sliced
2 garlic cloves, minced
1 tablespoon grated fresh ginger
1 (14-ounce) can whole peeled tomatoes
½ cup chopped cilantro, divided
½ cup heavy cream
¼ cup lemon juice

1. Preheat the water bath to 146°F.
2. Combine the cumin, paprika, coriander, garlic powder, ginger, turmeric, cayenne, and salt in a small bowl and mix well. Set aside 3 tablespoons of the spice mixture.
3. To bag the chicken, start by folding the tops of two vacuum-seal or zip-top bags back over themselves to form cuffs. This prevents chicken juices from getting on the edges of the bag, which can interfere with the seal or spread bacteria.
4. Add the chicken, the remaining spice mixture, and 2 tablespoons of butter, squeezing the bags to coat every surface. Uncuff and seal the bags using a vacuum sealer or the displacement method.
5. Place the bags in the water bath and cook for 1 hour (the chicken can stay in the bath up to 2 hours).

6 While the chicken is in the water bath, heat the remaining 3 tablespoons of butter in a large pan or Dutch oven over medium-high heat until melted and the foaming subsides.

7 Add the onion, garlic, and ginger. Cook, stirring frequently, for about 10 minutes.

8 Add the reserved 3 tablespoons of spice mixture and cook, stirring frequently, until fragrant, about 30 seconds.

9 Add the tomatoes and half the cilantro, scraping up any browned bits from the bottom of the pan with a spoon. Simmer for 15 minutes, then purée using an immersion blender, or transfer to a stand blender and blend until smooth.

10 Stir in the cream and lemon juice. Season with salt, then set aside until the chicken is cooked.

11 Remove the chicken bags from the water bath. Place each piece into the pan with the sauce and stir until fully coated. Stir in most of the remaining cilantro, keeping a small handful for garnish.

12 Plate and top with cilantro. Serve.

Tip: *If using the displacement method, spend a little extra time working out the air bubbles to reduce the possibility of the bags floating.*

Chicken Lettuce Wraps

YIELD: 4 SERVINGS / ACTIVE TIME: 20 MINUTES / TOTAL TIME: 20 MINUTES

PF Chang's chicken lettuce wraps are iconic. The good news is that you can make a similar dish right in your own kitchen, one that plays up the perfect flavors of succulent sous vide chicken with hoisin sauce, soy sauce, sriracha, ginger, onion, and water chestnuts. The combination is to die for and is also proof that you don't need to be a professional chef in order to cook delectable lettuce wraps that will leave everyone wanting a second serving.

- 1 tablespoon olive oil
- 1 (8-ounce) can whole water chestnuts
- 2 garlic cloves
- 1 small onion, diced small
- Cooked Chicken Breasts or Chicken Thighs (page 80 and page 82), minced
- 1 tablespoon ginger
- ¼ cup hoisin sauce
- 2 tablespoons soy sauce
- 1 tablespoon rice wine vinegar
- 1 tablespoon sriracha, plus more as needed
- 2 scallions, green and white parts thinly sliced
- Salt
- Freshly ground black pepper
- 1 head butter lettuce
- Chopped peanuts (optional)

1. Heat the oil in a saucepan over medium-high heat. While the oil is heating, chop the water chestnuts and set aside.
2. Add the garlic to the saucepan and stir until fragrant, about 30 seconds.
3. Add the onion and cook until translucent, 1 to 2 minutes.
4. Add the chicken, ginger, hoisin sauce, soy sauce, vinegar, and sriracha and cook, stirring, for 1 to 2 minutes.
5. Add the water chestnuts and scallions and cook until the scallions soften, 1 to 2 minutes. Season with salt and pepper.
6. To serve, separate the butter lettuce leaves and spoon in some chicken mixture.
7. Top with additional sriracha and peanuts (if using).

Duck Leg Confit

YIELD: 4 SERVINGS / ACTIVE TIME: 15 MINUTES / TOTAL TIME: 14 HOURS

Confit may take a while to prepare, but that's kind of the whole point. And it sure produces tasty results. This sous vide–style duck leg confit is fatty, flavorful, and definitely worth the effort. You don't need much in terms of spices or added flavors here, as the meat will speak for itself.

4 duck leg-thigh quarters
2 tablespoons kosher salt
½ cup rendered duck fat (or butter)
Freshly ground black pepper

1 To cure the duck legs, vacuum seal them with the salt. Start by folding the top of two vacuum-seal bags back over themselves to form a cuff. This prevents duck juices from getting on the edges of the bags, which can interfere with the seal or spread bacteria.

2 Add two duck legs to each of the bags, pack the salt as evenly as you can over the legs, uncuff, then seal using a vacuum sealer. Refrigerate for 6 hours.

3 Preheat the water bath to 180°F.

4 Remove the duck legs from the bags, rinse well with cold water, and pat the legs dry with paper towels. Transfer to new bags, divide the duck fat evenly between the bags, and vacuum seal the legs.

5 Place the bags in the water bath and cook for 8 hours.

6 Remove the duck legs from the bags, place on a paper towel–lined plate, and pat dry.

7 Heat a cast iron or stainless steel skillet on high and add the duck legs, skin-side down. Sear for 20 to 30 seconds on each side until golden brown.

8 Season with pepper (they've already brined in salt, so they're probably salty enough).

Tip: *I love to serve this with a pudding for dessert, since they cook at the same temperature. See Chapter 7 for ideas.*

Challenge: *Use duck or chicken gizzards in place of legs!*

Five-Spice Duck Breast with Blackberries and Sage

YIELD: 4 SERVINGS / ACTIVE TIME: 25 MINUTES / TOTAL TIME: 1 TO 2 HOURS

Duck meat packs more punch than other poultry, and its intense flavor profile and rich texture are enhanced when cooking it sous vide. The blackberry and sage topping complements its flavor even more. Sweet brown sugar, fresh sage, shallots, and blackberries make for a warm and herbaceous meal that's also just the right amount of sweet.

- 4 skin-on, boneless duck breasts
- Chinese Five Spice powder (see page 217)
- 1 tablespoon olive oil or duck fat
- 1 shallot, finely diced
- 8 fresh sage leaves
- ¼ cup sherry vinegar
- ¼ cup brown sugar
- 8 ounces fresh blackberries, divided
- 1 cup Chicken Stock (page 207)
- ¼ cup port (optional)

1. Preheat the water bath to 125°F for medium-rare or 135°F for medium.
2. Season the duck breasts with five-spice rub.
3. To bag the duck, start by folding the top of a vacuum-seal bag back over itself to form a cuff. This prevents duck juices from getting on the edges of the bag, which can interfere with the seal or spread bacteria. For best results, do not bag more than two pieces together in one bag.
4. Add the seasoned duck breasts to the bags. Uncuff and seal using a vacuum sealer.
5. Place the bags in the water bath for 1 to 2 hours.

6 When done, remove the duck breasts from the pouches to a paper towel–lined plate and pat dry.

7 Score the skin with a sharp knife at an angle about every half-inch, and then repeat at a 90-degree angle to form a diamond pattern to allow the fat to render more easily.

8 Heat the oil or fat in a large cast iron or stainless steel skillet over medium-high heat until the oil is shimmering.

9 Place the duck breasts, skin-side down, in the skillet and cook until the skin is crispy, 30 seconds to 1 minute. Flip the breasts and cook for 30 seconds more. Remove the duck from the pan and tent with foil to keep warm.

10 Reduce the heat to medium, and add the shallot and sage. Cook until softened, 2 to 3 minutes. Add the vinegar, brown sugar, and half the blackberries, stirring until the sugar is dissolved and the berries have released their juice.

11 Add the stock and port (if using) and increase the heat. Simmer until the liquid is reduced by half, 5 to 10 minutes.

12 Strain the contents of the skillet, discarding any solids.

13 To serve, slice the duck breast thinly across the grain. Spoon the sauce over the meat and garnish with the remaining blackberries.

Thanksgiving Turkey

YIELD: 8 SERVINGS / ACTIVE TIME: 30 MINUTES / TOTAL TIME: 6½ HOURS

Yes, you can cook an entire turkey sous vide! Break down your bird, have your butcher do it, or buy parts separately (with a whole bird, heat wouldn't be evenly distributed because of the air cavity). It's the perfect choice when everyone wants a drumstick! Start the dark meat at a higher temperature, cool the bath down, and then add the white meat. Season with your favorite rub or keep it simple with salt and pepper only. This year, there will be no dry turkey.

1 (14-pound) turkey, defrosted
Salt
Freshly ground black pepper
1 cup Napa Valley Rub (page 215, or your favorite in place of salt and pepper; see pages 214–219, for options)
½ cup (1 stick) unsalted butter, divided
4 or 5 sage sprigs
4 or 5 thyme sprigs
4 or 5 rosemary sprigs
Ice
2 tablespoons olive oil, divided

1 Preheat the water bath to 165°F.

2 If starting with a whole turkey, remove the packaged gizzards inside the body cavity (if present) and set them aside for gravy. Remove the thighs and drumsticks. Remove the wings. Cut out the rib cage and save it for stock or gravy. Separate the breasts from the bone.

3 Season the turkey pieces all over with salt and pepper or your favorite rub.

4 Prepare four 1-gallon vacuum-sealer bags or four zip-top bags. Place the thighs and wings in one, making sure they do not overlap, and add 1 sprig each of sage, thyme, and rosemary, then 2 tablespoons of butter. Repeat with the drumsticks in another bag with the herbs and butter. You may need two bags depending on the size of the drumsticks.

5 Place one turkey breast in each of the two remaining bags with 1 sprig each of sage, thyme, and rosemary, then add 2 tablespoons of butter to each bag.

6 Seal the bags using a vacuum sealer or the displacement method.

7 Set aside the breasts in the refrigerator while the thighs and drumsticks cook.

8 Place the thighs and drumsticks in the water bath and cook for 3½ hours.

9 Chill the water bath to 146°F by adding a few ice cubes.

10 Add the breasts to the water bath (do not remove the legs and thighs) and cook for an additional 2½ hours.

11 To serve immediately, remove the meat from the bags and pat the skin dry with paper towels.

12 Heat 2 teaspoons of oil in a stainless steel or cast iron pan on high heat until it shimmers. Add 2 turkey pieces, skin-side down, and sear until brown and crispy, about 3 minutes per side. Repeat with the remaining oil and turkey.

13 To store for later, immediately add the turkey to an ice bath after removing from the water bath and chill for at least 30 minutes before placing in the refrigerator.

14 To reheat, put the bags back in water bath at 146°F for 1 hour, then follow steps 11 and 12 to crisp up the skin.

Tip: *To make a complete meal sous vide, cook the vegetables first, then let them hang out in the water bath as you lower the temperature and cook the meats. Then everything will be done at once and can be served at the same temperature.*

Turkey Pumpkin Roulade with Apple Cider Gravy

YIELD: SERVES 4 / ACTIVE TIME: 30 MINUTES / TOTAL TIME: 2½ HOURS

Roulade comes from the French verb *rouler*, meaning "to roll." These crafty-looking dishes are as delicious as they are fun to look at. This recipe uses turkey breasts, pumpkin, pecans, apple cider, and olive oil for a delicious meal perfect for fall weather. The flavors and tender textures are out of this world. Don't be intimidated by the rolling process, either. It's not as difficult as it looks, and this turkey roulade is definitely worth the effort.

2 (1-pound) boneless, skinless turkey breasts
Salt
Freshly ground black pepper
1 cup canned pumpkin purée (not pumpkin pie filling)
¼ cup chopped toasted pecans
2 tablespoons crumbled sage, plus 2 teaspoons
1¼ cups chicken broth, divided
¼ cup apple cider
2 tablespoons olive oil
2 tablespoons flour

1 Preheat the water bath to 146°F.

2 Using a rolling pin, heavy skillet, or meat pounder, flatten out both breasts to about ¼ inch. Season each breast on both sides with salt and pepper.

3 In a small bowl, stir together the pumpkin purée, the pecans, 2 tablespoons of sage, and ⅛ teaspoon of salt.

4 Spread half the filling on each breast, leaving a ½-inch border around it. Roll up each breast jelly-roll style, starting at the narrow end, and secure every 1 to 2 inches with baker's twine.

5 To bag the roulades, start by folding the top of a zip-top bag back over itself to form a cuff. This prevents turkey juices from getting on the edges of the bag, which can interfere with the seal or spread bacteria.

6 Pour ¼ cup of chicken broth and the apple cider into the bag. Return the remaining chicken broth to the refrigerator. Add the turkey roulades, uncuff the bag, and seal using the displacement method.

7 Place the bag in the water bath and cook for 2 hours.

8 Once the turkey is fully cooked, remove the roulades from the bag, reserving the cooking liquid. Place them on a paper towel–lined plate and pat dry.

9 In a medium ovenproof skillet, heat the oil over a medium-high heat. Add the turkey and brown 1 to 2 minutes on each side. Transfer to a plate and keep warm.

10 In a measuring cup, vigorously whisk together the remaining 1 cup of chicken broth and the flour. Add the reserved cooking liquid and remaining 2 teaspoons of sage.

11 Add the liquid mixture to the skillet, bring to a boil, and then reduce to a simmer until thickened, 8 to 10 minutes. Season liberally with salt and pepper.

12 Cut the turkey into 1-inch slices and serve with the gravy.

Tip: *This dish is also great with pork! Use a butterflied pork tenderloin in place of turkey breasts and cook as instructed.*

Tender Pork Cuts: Tenderloin, Chops, Cutlets

YIELD: 1 TO 12 SERVINGS / ACTIVE TIME: 10 MINUTES
TOTAL TIME: 2 TO 5 HOURS, DEPENDING ON DESIRED DONENESS

Much like steaks, pork can be cooked to your preferred level of doneness—even rare! The following recipe is what I consider to be the ideal finish for pork—medium-rare.

- Tenderloin (1 to 1½ pounds) or chops (1 inch thick)
- Salt
- Freshly ground black pepper
- Your favorite rub (in place of salt and pepper)
- Rosemary sprigs (optional)
- Thyme sprigs (optional)
- 1 tablespoon olive oil
- 1 tablespoon butter

1 Preheat the water bath to the correct temperature for your desired doneness (I prefer 140°F).

2 Season the pork all over with salt and pepper or your favorite rub.

3 To bag the pork, start by folding the top of a vacuum-seal or zip-top bag back over itself to form a cuff. For best results, do not bag more than one tenderloin or two chops together in one bag. Add the pork and any aromatics, such as rosemary or thyme (if using). Uncuff and seal the bag(s) using either a vacuum sealer or the displacement method.

4 Place the bag(s) in the water bath and cook for 2 to 4 hours. Thinner chops and cutlets may need as little as 1½ hours.

5 Remove the pork from the bag(s) and place it on a paper towel–lined plate. Pat dry on all sides.

6 Heat the oil in a large cast iron or stainless steel skillet over medium-high heat until the oil is shimmering. Add the pork, and cook until the bottom has developed a nice brown crust, about 45 seconds. Turn the pork with your tongs.

7 Add the butter to the skillet and let it melt, spooning the melting butter over the pork. Cook for 30 seconds. Continue until all sides are browned.

8 Alternatively, sear the pork on a grill or with a torch, making sure to brown all sides.

9 Remove the pork from the heat. Slice and serve with the drippings from the pan.

MEDIUM-RARE	140°F	2 to 4 hours
MEDIUM	150°F	2 to 4 hours
MEDIUM-WELL	160°F	2 to 5 hours

Tough Pork Cuts: Butts and Shoulders

YIELD: 3 TO 5 POUNDS / ACTIVE TIME: 10 MINUTES / TOTAL TIME: 8 TO 24 HOURS, DEPENDING ON DESIRED DONENESS

Pork butts actually come from the top of the shoulder rather than the . . . er, butt. Whether you call it pork butt, Boston butt, or pork shoulder, it makes delicious pulled pork almost any way you cook it. But one thing you can do only with sous vide is turn this tough cut into steaks as succulent as prime pork chops. The texture varies based on time and temperature; I encourage you to experiment to find your favorite combination.

3 to 5 pounds pork butt or shoulder
2 tablespoons oil
Salt and freshly ground black pepper or ¾ cup Sweet and Spicy Rub (page 216) or your favorite rub

1. Preheat the water bath to 145°F.
2. Because these cuts are generally large and can be hard to sear after prolonged cooking times, I recommend pre-searing the pork. To do this, first place the pork on a paper towel–lined plate and pat dry on all sides.
3. Heat the oil in a large cast iron or stainless steel skillet over high heat until the oil is almost smoking. Add the pork and sear on all sides, about 2 minutes per side.
4. Remove the pork to a plate and season all over with salt and pepper or your favorite spice rub (if using). Since I like to use this pork in a variety of dishes, from carnitas to barbecue sandwiches, I tend to keep the spices to a minimum and add them later.

5 To bag the pork, start by folding the top of a vacuum-seal bag back over itself to form a cuff. For best results, do not use zip-top bags. Uncuff and seal using a vacuum sealer.

6 Place the bags in the water bath and cook for at least 12 hours for sliceable steaks and up to 48 hours for shredding.

7 Remove the pork from the bag and place it on a paper towel–lined plate and pat dry on both sides.

8 If you want to recrisp the crust, you can sear it again. Use tongs to turn the meat in the hot pan.

9 If you are shredding, remove the meat to a cutting board and use two forks or your hands to pull the meat apart.

Tip: *Sous vide can save you money. By using low temperatures and long cooking times, cheap cuts of meat like brisket or pork shoulder can be turned into tender, succulent steaks.*

MEDIUM-RARE	145°F	8 to 24 hours
MEDIUM	155°F	8 to 24 hours
MEDIUM-WELL	165°F	8 to 16 hours

Perfect Pork Chops with Caramelized Apples

YIELD: 4 SERVINGS / ACTIVE TIME: 30 MINUTES / TOTAL TIME: 2 TO 4 HOURS

Caramel apples call to mind autumn, with its crisp flavors and bountiful harvest. This sous vide pork chop recipe is no different. The delectable taste and textures of the pork perfectly complement the tart, sweet apples, which are sautéed in brown sugar, cinnamon, nutmeg, and butter. The result is a classic play of sweet and savory enhanced by the richness of sous vide–style meats.

FOR THE PORK

4 (1-inch-thick) pork chops
Salt
Freshly ground black pepper
Sweet Spice (page 215) or your favorite rub (to replace salt and pepper)
1 tablespoon canola oil
1 tablespoon butter

FOR THE APPLES

2 tablespoons brown sugar
⅛ teaspoon ground cinnamon
⅛ teaspoon ground nutmeg
Dash salt
Dash freshly ground black pepper
2 tablespoons unsalted butter
2 crisp apples, peeled, cored, and sliced

1 Preheat the water bath to 140°F.

2 Season the pork chops generously with salt and pepper or your favorite rub on all sides.

3 To bag the pork, start by folding the top of a vacuum-seal or zip-top bag back over itself to form a cuff. For best results, do not bag more than two chops in one bag. Add the chops to the bags. Uncuff and seal the bags using either a vacuum sealer or the displacement method.

4 Place the bags in the water bath and cook for 2 hours (the chops can stay in the bath up to 4 hours).

5 Remove the pork chops from the bags and place them on a paper towel–lined plate. Pat dry very carefully on both sides.

6 Heat the oil and butter in a heavy cast iron or stainless steel skillet over high heat. Swirl until the butter is melted and starting to brown. Carefully add the pork chops and brown for about 45 seconds on each side.

7 When the chops are browned, pick them up with a pair of tongs and make sure to brown the edges as well. Transfer to a plate and let rest.

8 Meanwhile, prepare the apples. In a bowl, mix together the brown sugar, cinnamon, nutmeg, salt, and pepper.

9 In the same skillet, add the butter and stir in the brown sugar mixture and apples. Cover and cook over medium heat until the apples are just tender, 5 to 7 minutes.

10 With a slotted spoon or spatula, transfer the apples to a plate.

11 Continue cooking the sauce, uncovered, until thickened slightly, 3 to 5 minutes. Spoon the sauce over the apples and chops and serve.

Tip: *Wash and reuse zip-top bags.*

Spiced Pork Tenderloin with Mango Salsa

YIELD: 4 SERVINGS / ACTIVE TIME: 20 MINUTES / TOTAL TIME: 2 HOURS, 20 MINUTES

Pork tenderloin is lean, light, and—well—tender, but that doesn't mean it can't have intense flavor. This recipe plays up the tastes and textures everyone loves about pork tenderloin while pairing it with a spicy mango salsa that gives it an extra burst of heat and sweetness. The play between pork and produce is nothing new, but this recipe takes it to a new level.

- ¼ cup light brown sugar
- ¼ cup Sweet and Spicy Rub (page 216)
- 2 (1-pound) pork tenderloins
- 2 tablespoons canola oil
- 2 mangoes, pitted, peeled, and finely diced
- 1 red bell pepper, finely diced
- ¼ cup chopped cilantro or parsley
- 3 tablespoons finely diced red onion
- 2 tablespoons freshly squeezed lime juice
- 1 small jalapeño, seeded and finely diced
- Salt
- Freshly ground black pepper

1. Preheat the water bath to 140°F.
2. In a medium bowl, whisk together brown sugar and spice mix. Rub the spice mixture over the tenderloins.
3. To bag the tenderloins, start by folding the top of a vacuum-seal or zip-top bag back over itself to form a cuff. Add the tenderloins. Uncuff and seal the bag using either a vacuum sealer on the moist setting or the displacement method.
4. Place the bag in the water bath and cook for 2 hours.
5. Remove the bag from the water bath. Remove the pork from the bag to a paper towel–lined plate and pat dry.
6. Heat the oil in a large skillet over medium-high heat until the oil is shimmering. Add the pork and sear until browned on all sides, 2 to 3 minutes total. Transfer to a plate and let rest.
7. Meanwhile, prepare the salsa. In a medium bowl, mix together the mangoes, bell pepper, cilantro, onion, lime juice, and jalapeño. Season with salt and pepper.
8. Slice the pork and serve topped with the mango salsa.

Hawaiian Pulled Pork Sliders

YIELD: 6 SERVINGS / ACTIVE TIME: 15 MINUTES / TOTAL TIME: 1 HOUR

Sliders are always fun, and you certainly can't argue with pulled pork. Add the right smoky-sweet flavors and a Hawaiian roll, and you're in business. A great way to use up leftover pork shoulder, these sliders are tastefully refined and bursting with flavor.

Cooked pork shoulder (page 104)
½ cup pineapple juice
2 teaspoons liquid smoke
¼ cup mayonnaise
2 tablespoons white wine vinegar
Half of a 14-ounce bag coleslaw mix
½ cup finely diced pineapple
1 thinly sliced scallion, green and white parts
Kosher salt
Freshly ground black pepper
12 Hawaiian rolls

1 Preheat the water bath to 140°F.

2 To reheat the pork, in a large zip-top bag add the pork, pineapple juice, and liquid smoke. Seal the bag using the displacement method.

3 Place the pork in the water bath and cook for 30 minutes to 1 hour until heated through.

4 In a large mixing bowl, whisk together the mayonnaise and vinegar. Stir in the coleslaw mix, pineapple, and scallions. Season with salt and pepper.

5 Split the rolls, mound each roll bottom with pulled pork and coleslaw, cover with the top half of the roll, and serve.

Chinese Crispy Roasted Pork Belly

YIELD: 6 SERVINGS / ACTIVE TIME: 30 MINUTES / TOTAL TIME: 6 TO 8 HOURS

This recipe comes from Sous Vide Life, the collaboration of two longtime friends, Trevin Chow and Adam Phillabaum. The blog is focused on creating high-quality meals with sous vide cooking techniques. Find more recipes at SousVideLife.com. In this tasty recipe, the trick to replicating true Chinese crispy roasted pork belly is twofold: First, you need to ensure the pork belly is as flat as possible when you're cooking it in your water bath. This recipe includes the use of wooden or metal skewers inside the bag, so be careful when sealing. Without some structural support during the sous vide process, the pork belly will naturally fold up, making it harder to crisp up in the oven in the final step. Second, a light brushing of white vinegar before you broil it will ensure a nice even, and faster, browning.

2 pounds skin-on pork belly
2 teaspoons salt
1 teaspoon Chinese Five Spice powder (page 217)
1 teaspoon white pepper
3 tablespoons white vinegar

1 Heat the water bath to 165°F.

2 Using a sharp paring knife, slide horizontal cuts through the skin of the pork belly about 1 inch apart. Be careful to cut only through the skin and not too deep into the underlying fatty layer.

3 Criss-cross two skewers (I use metal ones) horizontally through the center of the layer of meat from the bottom left corner to the top right corner and from the bottom right corner to the top left corner, making a large X. This keeps the pork belly flat as it cooks sous vide, which makes it easier to crisp the entire layer of skin later, under the broiler.

4 In a small bowl, combine the salt, Chinese Five Spice powder, and white pepper. Sprinkle half of the mixture generously to cover the underside of the pork belly (opposite side of the skin).

5 To bag the pork, start by folding the top of a vacuum-seal bag back over itself to form a cuff. Add the pork belly. Uncuff and carefully seal using a vacuum sealer. If the tips of your skewers are exposed on one side of the pork belly, make sure you don't puncture the bag as you vacuum seal or handle it.

6 Place the bag in the water bath and cook for 6 to 8 hours. Any longer than this will significantly change the texture.

7 At the end of the sous vide bath, preheat the broiler to its high setting. Cut open the bag, pour away the accumulated liquid, and remove the skewers.

8 Using paper towels, pat the entire pork belly completely dry, with extra focus on the skin. Make several passes to make sure it's completely dry.

9 Sprinkle the remaining spice mixture on the underside of the belly (opposite side of the skin).

10 Flip over the belly, and brush it with a very thin layer of white vinegar. This will help it crisp more quickly and evenly in the oven.

11 Lay the pork belly skin-side up on a rack over a baking sheet so it is elevated off the tray.

12 Put it under the broiler and cook until the skin crisps and it is a nice dark brown, 10 to 15 minutes. As it's cooking, you will need to move the tray around to avoid hot spots burning too much of the skin.

13 Place the pork belly on a cutting board, and using a large chef's knife or meat cleaver, cut it into bite-size pieces.

14 Although I leave the skin intact, you can scrape off any of the skin that's too blackened in order to improve the presentation. Serve with white rice and Chinese chili sauce.

Easy Pork Shoulder Carnitas

YIELD: 8 SERVINGS / ACTIVE TIME: 20 MINUTES / TOTAL TIME: 30 MINUTES OR 48 HOURS

Carnitas have an amazing juicy, shredded texture that's hard to beat—the spices that come into play (cumin, oregano, coriander, salt, and paprika) are just the tip of the iceberg. Fair warning: You'll probably never go back to Chipotle again. If you already have cooked pork shoulder, it's even easier—just shred, season, and crisp.

FOR THE PORK IF STARTING FROM SCRATCH

4 to 5 pounds pork butt or shoulder
4 tablespoons canola oil, divided
1 teaspoon ground cumin
1 teaspoon ground coriander
1 teaspoon dried oregano
1 teaspoon paprika
1 teaspoon salt
2 bay leaves

OR FOR THE PORK IF STARTING WITH PRECOOKED PORK

2 tablespoons canola oil
2 pounds unseasoned precooked, shredded pork butt or shoulder
½ teaspoon ground cumin
½ teaspoon ground coriander
½ teaspoon dried oregano
½ teaspoon paprika
½ teaspoon salt

FOR ASSEMBLY

1 white onion, chopped
1 bunch cilantro, chopped
16 corn tortillas
1 lime, cut into wedges
Salsa (optional)

TO MAKE THE PORK IF STARTING FROM SCRATCH

1 Preheat the water bath to 145°F.

2 Because these cuts are generally larger and can be harder to sear after prolonged cooking times, I recommend pre-searing the pork. To do this, place the pork on a paper towel–lined plate and pat dry on all sides.

3 Heat the oil in a large cast iron or stainless steel skillet over high heat until the oil is almost smoking. Add the pork and sear on all sides, about 2 minutes per side.

4 In a small bowl, mix together the cumin, coriander, oregano, paprika, and salt. Rub the shoulder with the spice mixture.

5 To bag the pork, start by folding the top of a vacuum-seal bag back over itself to form a cuff. Add the shoulder and the bay leaves. Uncuff and seal using a vacuum sealer.

6 Place the packet in the water bath and cook for 48 hours.

7 Remove the roast from the bath, transfer to a paper towel–lined plate, and pat dry.

8 Heat 2 tablespoons of oil in a cast iron or stainless steel skillet over high heat. Add the roast and sear again on all sides, about 2 minutes per side. Transfer to a plate, and using two forks, pull the pork apart.

9 Heat the remaining 2 tablespoons of oil in the skillet until sizzling hot.

10 Add the shredded pork in batches and fry until crisp.

11 Drain on paper towels.

TO MAKE THE PORK IF STARTING WITH PRECOOKED PORK

1 Heat the oil in a cast iron or stainless steel skillet over high heat. Add the shredded pork, cumin, coriander, oregano, paprika, and salt. Stir-fry until crisp.

2 Drain on paper towels.

TO ASSEMBLE

1 In a medium bowl, mix together the onion and cilantro.

2 To heat the tortillas, either wrap them in foil and heat for 15 minutes in a 350°F oven, or wrap them in a damp paper towel and microwave for 10 seconds, or carefully toast them over an open gas flame.

3 Stack two tortillas together, and top with pork and the onion and cilantro mixture.

4 Serve with lime wedges and salsa (if using).

Tip: *Cover and insulate the water bath with clean towels to reduce evaporation and help save energy.*

Ribs with Whiskey Barbecue Sauce

YIELD: 4 SERVINGS / ACTIVE TIME: 45 MINUTES
TOTAL TIME: 24 HOURS, 45 MINUTES

Barbecued ribs are among my favorite main dishes, and for a very good reason. Their smoky taste and tender, meaty consistency are a killer combination, triggering all of the fat- and flavor-loving instincts in our brains. When prepared sous vide style, they practically melt off the bone, adding a rich dimension that makes them completely irresistible. The long cooking time of this recipe is worth the wait, I promise. You can use store-bought barbecue sauce in a pinch.

FOR THE RIBS

2 racks pork ribs (back ribs or spareribs)
¾ cup Sweet and Spicy Rub (page 216)
1 teaspoon liquid smoke

FOR THE BARBECUE SAUCE

¼ cup Sweet and Spicy Rub (page 216)
1 medium onion, minced
1¼ cups ketchup
2 tablespoons dark molasses
2 tablespoons spicy mustard
2 tablespoons bourbon whiskey
2 tablespoons Worcestershire sauce
1 tablespoon apple cider vinegar
1 large garlic clove, minced
1 teaspoon liquid smoke

1 Preheat the water bath to 145°F.

2 Divide the rib racks into four portions.

3 Rub the ribs generously on all sides with ¾ cup of spice rub mixture.

4 To bag the pork, start by folding the tops of four vacuum-seal or zip-top bags back over themselves to form cuffs. Place individual portions of rubbed ribs and a few drops of liquid smoke in each bag. Uncuff and seal the bags using either a vacuum sealer or the displacement method. If using zip-top bags, I recommend double-bagging because of the long cook time.

5 Place the bags in the water bath and cover the bath with plastic wrap. Cook for 24 hours, checking periodically to flip the bags and top up the water levels as needed.

6 To make the sauce, in a medium saucepan, whisk together the spice rub, onion, ketchup, molasses, mustard, whiskey, Worcestershire sauce, vinegar, garlic, and liquid smoke.

7 Bring to a bare simmer and cook until reduced and thickened, about 20 minutes. Set aside.

8 To finish the ribs, preheat the oven to 400°F. Remove the ribs from the vacuum bags and carefully pat dry with paper towels.

9 Line two rimmed baking sheets with aluminum foil and place a wire rack on each. Divide the ribs evenly on the racks, facing up. Transfer the ribs to the oven and cook until the surface is sizzling, 10 to 15 minutes.

10 Brush the ribs with about a third of the sauce and return to the oven for 5 minutes.

11 Remove the ribs from the oven, brush with another layer of sauce, and return to the oven until the sauce is dried and the ribs are sticky, about 5 minutes longer.

12 Remove the ribs from the oven, brush with another layer of sauce, and serve.

Tip: *Finish the ribs on the grill instead of the oven for a super smoky taste!*

Sous Vide Sausage

YIELD: 8 SERVINGS / ACTIVE TIME: 10 MINUTES / TOTAL TIME: 1 HOUR

Sausages are so delicious, and it's hard to mess them up. They're fatty, meaty, and juicy, and the right amount of seasoning can really take them to the next level. While it's hard to make a bad sausage, that doesn't mean it's easy to make a *fantastic* sausage. Keeping them perfectly juicy and flavorful is a common challenge, and to accomplish this, I usually recommend the sous vide method for cooking.

8 raw natural-casing sausages such as Bratwurst or Italian
6 ounces beer (optional)
2 teaspoons salt (optional)
1 tablespoon vegetable oil or butter (for stove top cooking)
Whole grain mustard for serving

1 Preheat the water bath to 160°F.

2 To bag the sausages, start by folding the top of a vacuum-seal or zip-top bag back over itself to form a cuff. Add the sausages. Make sure to keep them in a single layer so that they don't overlap. If you're worried about them keeping their shape when you seal them, add the beer and salt (if using). The salt helps keep the beer from drawing all of the salt and flavor out of the sausages!

3 Uncuff and seal the bag using a vacuum sealer on the gentle setting (don't squish those links and draw out liquid if you're using beer) or the displacement method.

4 Place the bag in the water bath and cook the sausages in the water bath for 45 minutes to 1 hour (they can hang out in the bath for up to 4 hours, but after that they'll get mushy).

5 When the sausages are cooked, remove them from the bag. Dry them completely and place on a paper towel–lined plate.

6 If you're finishing your sausages on the stove top, heat the oil over medium heat in a cast iron or stainless steel skillet. If you're grilling, heat a gas grill to high or build a hot fire in a charcoal grill. Add the sausages. For both the stove and the grill, sear the sausages for about 3 minutes total, turning over as needed until browned. Serve immediately with your favorite whole grain mustard.

Momofuku-Inspired Bao
with Quick Pickles and Hoisin Sauce

YIELD: 12 SERVINGS / ACTIVE TIME: 30 MINUTES / TOTAL TIME: 30 MINUTES (PLUS TIME TO COOK THE PORK BELLY)

Bao hardly needs any introduction. Inspired by renowned chef David Chang's steamed filled buns, this recipe helps you make Momofuku-worthy bao at home using hoisin sauce, quick pickles, and pork belly. Though bao comes in many forms, it's always irresistibly soft, a perfect vessel for whatever flavorful meat it holds in its center—in this case, that meat happens to be tender pork. Bao buns can be a little difficult to make from scratch, but there are many premade options available in specialty stores.

24 frozen bao (usually 2 packages)
Hoisin sauce
Cooked pork belly (page 110)
Quick pickles (recipe follows)
Sliced scallion, green and white parts, for serving (optional)
Sriracha, for serving (optional)

FOR THE QUICK PICKLES

1 seedless cucumber, very thinly sliced
¼ cup seasoned rice vinegar
1 tablespoon sugar
Pinch salt

1 To heat the bao, follow package directions or wrap in foil and steam for about 10 minutes.

2 Open a warmed bun and spread about ½ teaspoon of hoisin sauce on the inside.

3 Add 2 pieces of pork belly, then top with 2 slices of pickle.

4 Add a scattering of scallions and a squirt of sriracha (if using). Repeat with the remaining buns, pork belly, and condiments.

TO MAKE THE QUICK PICKLES

In a medium bowl, stir together all the ingredients. Marinate in the refrigerator for at least 15 minutes or up to 4 hours.

Chashu Ramen with Mushrooms and Eggs

YIELD: 4 SERVINGS / ACTIVE TIME: 30 MINUTES / TOTAL TIME: 36 HOURS

This recipe comes from Jennifer Che ("Jen"), author of Tiny Urban Kitchen, a food and travel blog highlighting Jen's culinary adventures. Pork belly can be tricky to cook—it has quite a bit of fat and collagen and needs to be cooked for a long time before it becomes tender—but this recipe is worth it. Be aware that the higher the temperature at which you cook your meat, the more moisture it loses, and thus the tougher it becomes. Therefore, making good pork belly involves walking that fine line of finding the optimal temperature and time.

FOR THE CHASHU

½ cup soy sauce
1 cup mirin
1 cup sake
½ cup sugar
6 scallions, green and white parts thinly sliced
6 whole garlic cloves
2-inch piece ginger, sliced into matchsticks
2-pound boneless pork belly, rolled up

Tip: *Use a plastic container rather than a metal pot to better insulate the water bath and help save energy.*

FOR THE RAMEN AND EGGS

4 eggs
2 teaspoons olive oil
4 ounces shiitake mushrooms, sliced
4 scallions, thinly sliced on the diagonal, white and green parts separated
4 garlic cloves, minced
2-inch piece ginger, minced
2 tablespoons sesame oil
½ cup ponzu sauce
2 tablespoons soy sauce
2 tablespoons brown sugar
2 teaspoons rice vinegar
3 cups water
Salt
Freshly ground black pepper
1 pound fresh ramen noodles
½ pound chashu (pork belly)
½ cup blanched corn kernels
½ cup bean sprouts
2 teaspoons furikake (or toasted sesame seeds)

1 Preheat the water bath to 146°F.

2 In a small saucepan, heat the soy sauce, mirin, sake, and sugar until the sugar melts. Stir in the scallions, garlic, and ginger. Take off the heat and cool.

3 Add the marinade and pork belly to a vacuum-seal or zip-top bag and seal gently by using a vacuum sealer or the displacement method. If using a zip-top bag, I recommend double-bagging.

4 Place the bag into the water bath and cook for 36 hours.

5 When the pork is done cooking, prepare an ice bath.

6 Remove the bag from the water bath and chill in the ice bath for 20 to 30 minutes, or until the pork belly is cool to the touch. While the pork belly cools, prepare the eggs and ramen (see step 8). Once cool, remove the pork belly from the bag, dry off, and cut into thin slices. You'll need about ½ pound for the ramen; refrigerate the rest and use in bánh mì or other sandwiches.

7 After removing the pork belly, add enough boiling water to raise the temperature of the water bath to 167°F.

8 When the water bath reaches temperature, carefully add the eggs and cook for 20 minutes.

9 Remove the eggs from the water bath and chill in an ice bath. When cool to the touch, peel the eggs.

10 While the eggs cook, in a large, high-sided pan (or pot), heat the olive oil over medium-high heat until hot. Add the mushrooms and cook, stirring occasionally, until browned and crispy, 4 to 6 minutes.

11 Add the white bottoms of the scallions, garlic, ginger, and sesame oil and cook, stirring occasionally, until fragrant, 30 seconds to 1 minute. Add the ponzu sauce, soy sauce, brown sugar, vinegar, and water.

12 Bring the mushroom mixture to a simmer and cook, stirring occasionally, until thoroughly combined, 4 to 6 minutes. Remove from the heat and season with salt and pepper.

13 Bring a separate pot of water to a boil and add the noodles, stirring gently to separate.

14 Cook the noodles until tender, 2 to 3 minutes. Drain thoroughly and rinse under warm water to prevent sticking.

15 Divide the drained noodles and mushroom broth among four bowls.

16 Cut the peeled eggs in half; season with salt and pepper. Top the noodles with the seasoned eggs and sliced pork belly. Garnish with the green tops of the scallions, corn kernels, bean sprouts, and furikake.

5

SEAFOOD

Seafood is absurdly easy . . . to overcook. That's why many people avoid cooking it altogether. However, the precision of sous vide makes cooking seafood much easier and more enjoyable; plus overcooking is much less likely. Salmon retains a color and juiciness that may be completely foreign to you. The same goes for shellfish and other challenging seafood (imagine: tender octopus). Because many fish are very delicate, adding oil to the bag to act as a cushion and prevent your food from sticking to the bag itself is a good idea. I recommend olive oil with seafood since we don't need the high smoke point that we want with meat. Ideal temperatures for fish and shellfish range from 122°F to 132°F. Experiment to find your preferred doneness.

Tender and Flaky Salmon

YIELD: 1 TO 6 FILLETS / ACTIVE TIME: 5 MINUTES / TOTAL TIME: 30 MINUTES TO 1 HOUR, DEPENDING ON THICKNESS

This is the simplest way to prepare salmon and let all its natural flavors shine through. Never again worry about serving salmon that's still too pink in the middle or overcooked and white. This method ensures that your fish will be both moist and flaky every time.

2 tablespoons olive oil, divided
4 salmon fillets, skinless or skin-on
Seasonings and garnishes of choice

1 Preheat the water bath to 122°F.

2 To bag the fillets, start by folding the top of a zip-top bag back over itself to form a cuff. Add 1 tablespoon of oil, then the salmon fillets. Uncuff and seal the bag using the displacement method. Make sure to keep the fillets in a single layer so that they don't overlap.

3 Place the bag in the water bath. Use the chart below to cook the fillet for the necessary amount of time.

4 Remove the salmon from the bag.

5 At this point, you can plate the salmon, and season with high-quality olive oil, flaky sea salt, or any other sauces or garnishes you prefer.

6 If your salmon is skin-on, you'll probably want that skin to be crisp. If so, you'll need to sear the fillet. Heat the remaining 1 tablespoon of oil in a large stainless steel skillet over medium-high heat until the oil is slightly shimmering.

7 Add the salmon skin-side down and cook for about 45 seconds, swirling the oil or pressing down on the salmon to ensure the most contact with the hot oil for maximum browning.

8 Remove the salmon from the heat and serve.

Tip: *When searing, ensure that the oil is very hot so that the time in the pan is kept to a minimum, between 30 seconds and 1 minute. Pat the food dry before searing, and sear in batches so as to not overly cool the pan.*

½-INCH-THICK FILLET	30 minutes
1-INCH-THICK FILLET	40 minutes
1½-INCH-THICK FILLET	1 hour

Perfect Shrimp with Cocktail Sauce

YIELD: 4 SERVINGS / ACTIVE TIME: 10 MINUTES / TOTAL TIME: 1 HOUR, 15 MINUTES

Since shrimp have so little fat, even a minute too long on the heat can turn them from moist and succulent to tough and rubbery. Dialing in a precise temperature ensures that you're cooking them to their ideal doneness without any guesswork—even if they're frozen.

1 pound shrimp, peeled and deveined
1 tablespoon butter or extra-virgin olive oil

COCKTAIL SAUCE
1 cup ketchup
¼ cup prepared horseradish
2 teaspoons Worcestershire sauce
1 teaspoon lemon juice
1 to 3 dashes hot sauce (like Tabasco)
Salt
Freshly ground pepper

Note: *Most shellfish tastes best cooked at 122°F; however, unless it is cooked for over 45 minutes, it is not fully pasteurized and should not be eaten if you have a weak immune system or the shellfish is not of sushi-grade quality. For fully pasteurized shellfish, it is best to cook them to 140°F or for a longer period of time. Be aware, though, this will not kill any viruses that are in the shellfish, but this is an issue with traditional cooking methods as well.*

1 Preheat the water bath to 130°F.

2 To bag the shrimp, start by folding the top of a vacuum-seal or zip-top bag back over itself to form a cuff. Add the shrimp and butter or oil. Uncuff and seal the bag using a vacuum sealer on the moist setting or the displacement method. Make sure to keep the shrimp in a single layer so that they don't overlap.

3 Place the bag in the water bath and cook for 15 minutes.

4 While the shrimp are cooking, prepare an ice bath.

5 Remove the bag from the water bath and transfer to the ice bath. When the shrimp are cooled to room temperature, transfer to the refrigerator.

6 Refrigerate for at least 1 hour before serving with cocktail sauce. You can also use the shrimp for pad Thai, scampi, or linguine.

TO MAKE THE COCKTAIL SAUCE
Combine all the ingredients in a small bowl and adjust the spices.

Thai-Style Halibut with Coconut-Curry Broth

YIELD: 4 SERVINGS / ACTIVE TIME: 20 MINUTES / TOTAL TIME: 40 MINUTES

Thai-style fish curry is healthy and wildly flavorful, making it a go-to choice for fish lovers everywhere. This recipe combines the sweet yet mild flavor of halibut with rich coconut curry steeped in lemongrass, garlic, and cilantro. Spinach and brown rice round out the recipe, making it nutrient-rich and full of complex flavor.

4 (6-ounce) pieces halibut fillet, skin removed
Salt
Freshly ground black pepper
3 tablespoons olive oil, divided
4 shallots, minced
2½ teaspoons red curry paste
2 cups Chicken Stock (page 207)
½ cup light coconut milk
1 lemongrass stalk, cut into 2-inch lengths (optional)
2 garlic cloves, minced
5 cups spinach
½ cup coarsely chopped fresh cilantro leaves
2 scallions, green parts only, thinly sliced
2 tablespoons freshly squeezed lime juice
2 cups cooked brown rice, for serving

1 Preheat the water bath to 122°F.

2 Season the halibut fillets with salt and pepper. To bag the fillets, start by folding the top of a vacuum-seal or zip-top bag back over itself to form a cuff. Add the fillets and 1 tablespoon of olive oil. Uncuff and seal the bag using a vacuum sealer on the moist setting or the displacement method. Make sure to keep the fillets in a single layer so that they don't overlap.

3 Place the bag in the water bath and cook for 20 minutes.

4 In a large sauté pan, heat 1 tablespoon of olive oil over medium heat until shimmering. Add the shallots and cook, stirring frequently, until beginning to brown, 3 to 5 minutes.

5 Add the curry paste and stir until fragrant, about 30 seconds.

6 Add the stock, coconut milk, lemongrass (if using), and ½ teaspoon of salt and simmer until reduced to 2 cups, about 5 minutes. Set aside.

7 In a stainless steel or cast iron pan, heat the remaining 1 tablespoon of olive oil over medium-high heat until shimmering. Add the garlic and cook until fragrant, about 30 seconds. Pile in the spinach and toss until wilted. Season with salt and pepper.

8 Arrange the spinach in the bottom of 4 soup bowls. Top with the fish fillets. Stir the cilantro, scallions, and lime juice into the curry sauce and season with salt and pepper. Ladle the sauce over the fish and serve with rice.

Tip: *If you can't find halibut, this is a great recipe for tilapia or cod.*

Nobu-Inspired Miso Black Cod

YIELD: 4 SERVINGS / ACTIVE TIME: 15 MINUTES / TOTAL TIME: 45 MINUTES

This is my take on the classic and elegant Japanese dish made famous by chef Nobu Matsuhisa. Traditionally steeped for days in a salty-sweet miso marinade, the buttery cod becomes deeply seasoned and completely transformed. The sous vide process speeds up the infusion of flavor by cooking the fish with the broth, so you can get similar results in a fraction of the time (though for mind-blowing flavor, you can let it marinate a few days in advance).

¼ cup cooking sake
¼ cup mirin
¼ cup miso paste
3 tablespoons brown sugar
4 (6-ounce) skinless black cod fillets
2 tablespoons olive oil
2 scallions, thinly sliced, green and white parts separated

1 Preheat the water bath to 130°F.

2 In a medium bowl, mix the sake, mirin, miso, white bottoms of the scallions, and brown sugar. Add the cod and toss to coat.

3 To bag the cod, start by folding the top of a zip-top bag back over itself to form a cuff. Add the cod and miso mixture. Uncuff and seal the bag using the displacement method. Make sure to keep the fillets in a single layer so that they don't overlap. At this point you can cook the fish or let it marinate in the refrigerator for up to 2 days.

4 Place the bag in the water bath and cook for 30 minutes.

5 Remove the cod from the bag and reserve the liquid.

6 Heat the oil in a large nonstick skillet over high heat. When the oil just begins to smoke, add the cod and sear until slightly blackened on one side, about 30 seconds. Transfer to a plate.

7 Return the skillet to medium-high heat and add the reserved cooking liquid. Bring to a rapid simmer and cook, stirring frequently, until reduced to a thick sauce, 5 to 10 minutes.

8 Arrange the cod on serving plates. Drizzle with the reduced sauce and garnish with the green parts of the scallions. Serve.

Tip: *If you can't find black cod, sustainably caught Chilean sea bass is a gorgeous substitution.*

Salmon and Lemon Dill Sauce

YIELD: 4 SERVINGS / ACTIVE TIME: 20 MINUTES / TOTAL TIME: 1½ HOURS

The simplest way to prepare salmon sous vide is with some olive oil, but in my house, we swear by brining the salmon first. This recipe uses a very simple salt-water brine to bring out the juiciness of the fish. You can add other seasonings to the brine, but only the salt will penetrate the protein so it all washes away in the end. Keep it simple and add the flavors at the end.

2 cups ice water
2 tablespoons kosher salt, plus more as needed
4 skin-on salmon fillets
2 tablespoons olive oil, divided
4 tablespoons Greek yogurt or sour cream
2 lemons, 1 juiced, 1 cut into quarters
¼ cup chopped fresh dill, plus 1 tablespoon
Freshly ground black pepper

1 To prepare the brine, in a zip-top bag, add the ice water and 2 tablespoons of salt and shake to mix.

2 Add the salmon fillets, push out as much air as possible, and seal the bag. Refrigerate for about 30 minutes.

3 Preheat the water bath to 122°F.

4 Remove the salmon fillets from the brine and pat dry with a paper towel. Discard the brine.

5 To bag the fillets, start by folding the top of a zip-top bag back over itself to form a cuff. Add 1 tablespoon of oil, then the salmon fillets. Uncuff and seal the bag using the displacement method. Make sure to keep the fillets in a single layer so that they don't overlap.

6 Add the bag to the water bath and cook for 1 hour.

7 Meanwhile, prepare the dill sauce: Stir together the yogurt, the juice from 1 lemon, and ¼ cup of dill until well mixed. Season with salt and pepper. Refrigerate until the salmon is prepared.

8 Remove the salmon fillets from the bag and place on a paper towel–lined plate and pat dry.

9 Heat the remaining 1 tablespoon of oil in a large skillet over medium-high heat until shimmering. Add the fillets skin-side down and sear to crisp, about 45 seconds.

10 Remove the salmon from the heat and serve with the lemon dill sauce, the lemon quarters, and the remaining 1 tablespoon of dill for garnish.

Twice-Cooked Scallops with Pancetta and Asparagus

YIELD: 4 SERVINGS / ACTIVE TIME: 40 MINUTES / TOTAL TIME: 50 MINUTES (1 HOUR, 20 MINUTES IF USING FROZEN SCALLOPS)

Because scallops are so delicate and have so little fat, they are quickly and easily overcooked, turning them into inedible rubber discs. Cooking them sous vide means they come out tender, buttery, and rich. Every. Single. Time. Serve them as is, on a salad, in paella, or any other way you can imagine. Even frozen scallops are completely transformed by this technique!

- 12 sea scallops, fresh or frozen
- 3 tablespoons butter, divided
- 12 medium asparagus spears, trimmed and sliced into 2½-inch pieces
- 3 teaspoons olive oil, divided
- ¼ teaspoon salt, plus more as needed
- ¼ teaspoon freshly ground black pepper, plus more as needed
- ⅔ cup diced uncooked pancetta or center-cut bacon (about 3½ ounces)
- ¾ cup freshly squeezed orange juice
- 2 tablespoons dry vermouth or dry white wine (optional)
- 2 tablespoons chopped chives
- 2 tablespoons shaved Parmesan cheese

1 Preheat the water bath to 122°F. Preheat the oven to 475°F.

2 To bag the scallops, start by folding the top of a vacuum-seal or zip-top bag back over itself to form a cuff. Add the scallops and 2 tablespoons of butter. Uncuff and seal the bag using a vacuum sealer on the gentle setting or the displacement method. Make sure to keep the scallops in a single layer so that they don't overlap.

3 Place the bag in the water bath and cook for 20 minutes (add an extra 30 minutes if cooking from frozen).

4 Meanwhile, on a baking pan, toss the asparagus with 1 teaspoon of olive oil, ¼ teaspoon of salt, and ¼ teaspoon of pepper. Bake the asparagus in the oven for 6 minutes. Remove the asparagus from the pan and set aside.

5 In a large skillet, cook the pancetta or bacon until crisp, about 3 minutes. Remove from the skillet and drain on paper towels. Reserve the drippings.

6 Prepare an ice bath. When the scallops are done, immediately transfer the bag to the ice bath to chill for 5 to 10 minutes. This step helps set a custard-like interior and prevents the scallops from dumping all their liquid when you sear them.

7 Remove the scallops from the bag, reserving the juices. Sprinkle both sides of the scallops with salt and pepper.

8 Heat the reserved pancetta drippings and remaining 2 teaspoons of olive oil in the skillet over medium-high heat until just smoking. Add the scallops. Sear quickly on both sides—no more than 15 to 30 seconds or they'll overcook. Set aside and keep warm.

9 Keep the skillet over medium-high heat, and add the reserved juices from the bag, the orange juice, and the vermouth (if using). Bring to a boil and cook, stirring constantly, until the mixture is reduced to ½ cup, 5 to 10 minutes. Remove from the heat, and whisk in the remaining 1 tablespoon of butter.

10 Place 3 scallops on each plate and divide the asparagus among the plates. Drizzle with the sauce; sprinkle with pancetta, chives, and Parmesan cheese; and serve.

Spanish-Style Octopus with Chorizo and Butter Beans

YIELD: 4 SERVINGS / ACTIVE TIME: 20 MINUTES
TOTAL TIME: 5 HOURS, 20 MINUTES

Octopus wasn't for me until I had it sous vide—what a transformation! Tough, rubbery tentacles were replaced by buttery, tender morsels more akin to scallops than any octopus I'd ever had. My entire world was rocked, and I was encouraged to try it out at home. You can buy octopus frozen, and it already comes cleaned. But if you opt to buy fresh, I recommend looking for just the tentacles or baby octopi (otherwise you'd need to clean it yourself).

FOR THE OCTOPUS

2 pounds small octopus tentacles (1 inch or less in diameter) or baby octopus
1½ tablespoons olive oil, plus more for searing
1 thyme sprig
1 rosemary sprig
2 bay leaves
¼ teaspoon coriander seeds
¼ teaspoon cumin seeds
1 dried red chile
Salt
Freshly ground black pepper

FOR THE BEANS

1 (14-ounce) can butter beans with their liquid
½ pound Spanish chorizo, chopped
3 tablespoons olive oil, divided
1 tablespoon sherry vinegar
Salt
Freshly ground black pepper
1 pint cherry tomatoes, halved
½ small red onion, thinly sliced
¾ cup coarsely chopped flat-leaf parsley, plus a few tablespoons of whole leaves for garnish

1 Preheat the water bath to 170°F.

2 To bag the octopus, start by folding the top of a vacuum-seal or zip-top bag back over itself to form a cuff. Add the octopus, 1½ tablespoons of olive oil, the thyme, the rosemary, the bay leaves, the coriander, the cumin, and the chile. Sprinkle with salt and pepper. Uncuff and seal the bag using a vacuum sealer on the gentle setting or the displacement method.

3 Place the bag in the water bath and cook for at least 5 hours (the octopus can stay in the water bath up for to 7 hours).

4 Let the octopus cool enough in the bag so you can handle it, then drain well. (The liquid can be reserved and even frozen for seafood stock later.)

5 While the octopus cools, prepare the beans and chorizo. In a small saucepan, warm the butter beans in their liquid over medium-low heat, 5 to 7 minutes.

6 In a large skillet, cook the chorizo over high heat until it is lightly browned and most of the fat is rendered, about 2 minutes.

7 In a medium serving bowl, whisk together 2 tablespoons of oil and the vinegar.

8 Stir in the tomatoes, onion, and ¾ cup of parsley.

9 Drain the beans and chorizo and add them to the bowl. Season with salt and pepper.

10 To finish the octopus, trim the tentacles and scrape away the top flesh with a sharp paring knife. Scrape away the membranes, including the suckers. If the suckers are difficult to remove, use your hands. You will be left with very clean, smooth pieces of octopus. (This step is optional.)

11 If you are using baby octopus, gently rub away the membrane—it will come off very easily.

12 Heat 1 tablespoon of olive oil in a stainless steel or cast iron skillet over high heat until smoking. Add the octopus and sear until the edges are crisp, about 2 minutes.

13 Transfer the octopus to a cutting board and cut into 2-inch pieces. To serve, divide the beans and chorizo among four plates, placing the tentacles on top. Garnish with the parsley leaves.

Swordfish with Avocado and Mango Salsa

YIELD: 4 SERVINGS / ACTIVE TIME: 15 MINUTES / TOTAL TIME: 45 MINUTES

Swordfish has a reputation for being meaty and steak-like, but when cooked sous vide it becomes so tender it's almost like butterfish. This mango salsa and avocado topping plays up that buttery, tender texture with flavor notes of thyme, red onion, serrano pepper, lime, and cilantro. The result is a fine-tuned seafood dinner that is relatively easy to make—but it sure doesn't taste like it.

FOR THE FISH

4 (6-ounce) swordfish steaks
Salt
Freshly ground black pepper
3 tablespoons olive oil, divided
8 thyme sprigs, divided
Zest and juice of 4 lemons, plus 1 lemon cut into wedges

FOR THE SALSA

1 avocado, diced
1 mango, chopped
1 tablespoon finely diced red onion
2 tablespoons chopped cilantro or parsley, divided
1 tablespoon minced red bell pepper
1 teaspoon minced serrano pepper
1 tablespoon sriracha (optional)
1 tablespoon fresh lime juice
Salt
Freshly ground black pepper

1 Preheat the water bath to 130°F.

2 Season the swordfish with salt and pepper. To bag the swordfish steaks, start by folding the top of two vacuum-seal or zip-top bags back over themselves to form cuffs. Place 2 steaks in each bag. Add 1 tablespoon of oil, 4 thyme sprigs, and the zest and juice from 2 lemons to each of the bags. Uncuff and seal the bags using a vacuum sealer on the moist setting or the displacement method.

3 Place the bags in the water bath and cook for 30 to 45 minutes.

4 Meanwhile, make the salsa. In a large bowl, combine the avocado, mango, onion, 1½ tablespoons of cilantro, bell pepper, serrano pepper, sriracha (if using), and lime juice. Season with salt and pepper.

5 To finish the fish, remove the bags from the water bath when cooking is completed.

6 Carefully remove the swordfish from the bags and pat very dry with paper towels.

7 Heat a grill, grill pan, or cast iron skillet with the remaining 1 tablespoon of olive oil on high until smoking. Add the swordfish and sear for 1 to 2 minutes on each side.

8 Plate the swordfish, top with the salsa, and sprinkle with the remaining ½ tablespoon of cilantro. Serve with the lemon wedges.

Pro tip: *This recipe is also great for tuna or mahimahi!*

Salmon Burgers with Aïoli and Arugula

YIELD: 4 SERVINGS / ACTIVE TIME: 20 MINUTES / TOTAL TIME: 1 HOUR

Aïoli is an all-star condiment, full of fatty flavor and lemony tang. This popular condiment is perfect for pairing with almost any food, but it's particularly great with salmon burgers. This recipe plays up the flaky, flavor-rich salmon with Dijon mustard, garlic, dill, olive oil, arugula, and, of course, aïoli. Once you hop on board the aïoli train, you may never want to get off.

4 (6-ounce) fillets skinless salmon
7 tablespoons plus 1 teaspoon olive oil, divided
2 garlic cloves
2 tablespoons mayonnaise
1 teaspoon lemon juice
Salt
Freshly ground black pepper
2 teaspoons Dijon mustard
½ cup panko bread crumbs
2 shallots, minced
2 tablespoons finely chopped dill
1 tablespoon drained capers
1 tablespoon butter
4 hamburger buns
Tabasco sauce to taste
1 cup arugula

1 Preheat the water bath to 122°F.

2 To bag the fillets, start by folding the tops of four large zip-top bags back over themselves to form cuffs. Add 1 salmon fillet and 1 tablespoon of oil to each of the bags. Uncuff and seal the bags using the displacement method. Bagging each fillet individually will ensure that they don't become stuck together during cooking.

3 Place the bags in the water bath and cook for 40 minutes.

4 While the salmon cooks, make the aïoli. Mince the garlic, then using the side of your knife, smash until it resembles a paste (or use a Microplane zester). In a small bowl, combine the mayonnaise, half the garlic paste, the lemon juice, and 1 teaspoon of olive oil. Stir to combine and season with salt and pepper.

5 When the salmon is done, remove it from the bags and pat dry.

6 Using a fork, flake the cooked salmon into small pieces and place in a bowl.

7 Add the remaining half of the garlic paste, the mustard, bread crumbs, shallots, dill, capers, and 1 tablespoon of oil to the salmon in the bowl. Season with salt and pepper. Gently mix to combine, then form into four ½-inch-thick burgers.

8 In a stainless steel or nonstick pan, heat 2 tablespoons of olive oil on medium-high heat until shimmering.

9 Add the salmon burgers and cook until browned, 1 to 2 minutes per side. Transfer to a paper towel–lined plate. Wipe out the pan with a paper towel.

10 In the pan used to cook the burgers, melt the butter over medium heat until hot.

11 Add the buns, cut-sides down, and toast until lightly browned, 30 seconds to 1 minute. Transfer to a clean, dry work surface. Spread a layer of the aïoli onto the cut sides of the toasted buns. Place a cooked burger onto each bun bottom and season with salt, pepper, and Tabasco. Top with the arugula and the toasted bun tops.

Pad Thai with Shrimp

YIELD: 4 SERVINGS / ACTIVE TIME: 20 MINUTES / TOTAL TIME: 40 MINUTES

Pad Thai, to me, is classic comfort food. I had thought it was too complicated to make at home, but with a few staples and extras (fish sauce lasts forever and is great in many dishes, and don't sweat the tamarind paste if you can't find it), I found myself less than an hour away from a plate of warm, sweet, chewy noodles.

1 pound jumbo shrimp, peeled and deveined, tails left on
1 tablespoon olive oil
8 ounces flat rice noodles
2 tablespoons vegetable oil
1 garlic clove, minced
2 large eggs
1½ tablespoons soy sauce
2 tablespoons freshly squeezed lime juice
2 tablespoons brown sugar
2 tablespoons tamarind paste (optional)
1 teaspoon fish sauce
Dash red pepper flakes
1 pound Chicken Breasts, cut into bite-size pieces (page 80; optional)
3 scallions, green and white parts sliced
¼ bunch fresh cilantro, roughly chopped
¼ cup chopped unsalted peanuts

1 Preheat the water bath to 135°F.

2 To bag the shrimp, start by folding the top of a vacuum-seal or zip-top bag back over itself to form a cuff. Add the shrimp and olive oil. Uncuff and seal the bag using a vacuum sealer on the moist setting or the displacement method. Make sure to keep the shrimp in a single layer so that they don't overlap.

3 Place the bag in the water bath and cook for 30 minutes.

4 While the shrimp cook, soak the noodles in room-temperature water for 20 to 30 minutes. When softened, drain the noodles.

5 Remove the shrimp from the bag and pat dry.

6 In a large skillet, heat the vegetable oil over medium heat. Add the garlic and cook until fragrant, 1 to 2 minutes.

7 Whisk the eggs lightly with a fork. Add to the skillet and cook just until they are lightly scrambled. Remove the skillet from the heat and set aside.

8 In a small bowl, stir together the soy sauce, lime juice, sugar, tamarind paste (if using), fish sauce, and red pepper flakes. Put the skillet with the scrambled eggs back on medium heat, and pour the sauce into the skillet. Add the noodles, shrimp, and chicken (if using), and toss to coat in the sauce and cook until the mixture is warm.

9 Sprinkle with the scallions, cilantro, and peanuts. Toss lightly to combine. Serve warm.

Linguine with Shrimp Scampi

YIELD: 4 SERVINGS / ACTIVE TIME: 20 MINUTES / TOTAL TIME: 40 MINUTES

Linguine is the perfect pasta to accompany shrimp. This is why you'll commonly see the two paired together at upscale grills and Italian bistros—the combination is simply delicious. This shrimp scampi recipe calls for linguine, jumbo shrimp, garlic, parsley, butter, and lemon zest. It's just light enough to enjoy all summer long but rich enough to feel deliciously indulgent.

1 pound jumbo shrimp, peeled and deveined, with tails left on
3 tablespoons olive oil, divided
1 tablespoon kosher salt, plus 1½ teaspoons
¾ pound linguine
3 tablespoons butter
4 garlic cloves, minced
¼ teaspoon freshly ground black pepper
⅓ cup chopped fresh parsley leaves
¼ cup lemon juice (about 2 lemons)
Zest of ½ lemon
1 lemon, thinly sliced
Red pepper flakes (optional)

1 Preheat the water bath to 135°F.

2 To bag the shrimp, start by folding the top of a vacuum-seal or zip-top bag back over itself to form a cuff. Add the shrimp and 1 tablespoon of olive oil. Uncuff and seal the bag with a vacuum sealer on the moist setting or using the displacement method. Make sure to keep the shrimp in a single layer so that they don't overlap.

3 Place the bag in the water bath and cook for 30 minutes.

4 Heat a large pot of water to a boil. Add 1 tablespoon of salt and the linguine, and cook for 7 to 10 minutes, or according to package directions.

5 Meanwhile, in a stainless steel or cast iron skillet, melt the butter and remaining 2 tablespoons of olive oil over medium-low heat. Add the garlic and sauté until fragrant, 30 seconds.

6 Add the shrimp, the remaining 1½ teaspoons of salt, and the pepper and sauté for another 30 seconds.

7 Remove from the heat.

8 Add the parsley, lemon juice, lemon zest, lemon slices, and red pepper flakes (if using), and toss to combine.

9 When the linguine is cooked, drain, and add to the shrimp and sauce in the skillet. Toss well and serve.

Quick Gravlax

YIELD: 8 SERVINGS / ACTIVE TIME: 10 MINUTES / TOTAL TIME: 1½ HOURS

For those not familiar with Nordic cuisine, it's rich in flavorful fats, fresh seafood, and a host of cured meats and fish. Recipes for gravlax, delicious cured salmon often served as an appetizer, bring out everything you love about that fish, but made the traditional way, it takes several days. With sous vide, you can cut that time down to a couple of hours. All you need is a few ingredients, and it's easy to cure this Nordic favorite at home.

4 tablespoons salt
4 tablespoons sugar
1 pound center-cut salmon fillet
1 teaspoon liquid smoke (optional)

1. Preheat the water bath to 115°F.
2. Mix together the salt and sugar, and completely cover the salmon in the dry cure mixture. Let sit for 30 minutes, then rinse.
3. To bag the fillet, start by folding the top of a vacuum-seal or a zip-top bag back over itself to form a cuff. Add the salmon fillet to a vacuum-seal or zip-top bag. Add a few dashes of liquid smoke (if using). Uncuff and seal the bag with a vacuum sealer on gentle setting or using the displacement method.
4. Place the bag in the water bath and cook for 30 minutes.
5. Prepare an ice bath. Transfer the bag to an ice bath for 15 to 20 minutes.
6. Use a carving knife to cut the salmon in half lengthwise.
7. Trim off any dark flesh from the skin side of each piece and slice on the bias as thinly as possible.
8. Serve as an appetizer or with toasted bagels, cream cheese, onions, and capers for a traditional breakfast.

6

VEGGIES AND SIDES

Sealing your vegetables in a bag before cooking them in a water bath helps retain natural plant sugars that would otherwise be diluted in a large volume of cooking water. That means brighter color, more concentrated flavor, and higher vitamin content. Vegetables contain pectin—a kind of glue that holds their cells together and keeps them firm. Pectin doesn't break down until 183°F, so no matter what veggies you're cooking, they won't become tender below 183°F. Keep this in mind when adding any vegetables to your meat dishes while cooking. They may lose some flavor to the bag at lower temperatures, but they will not break down enough to cook. All vegetables will cook at the same temperature (183°F), but you can experiment with higher temperatures (up to 190°F) and lower cook times for the more tender leafy greens.

Tender Green Vegetables

YIELD: 6 TO 8 SERVINGS / ACTIVE TIME: 10 MINUTES / TOTAL TIME: 1 HOUR, 10 MINUTES

Cooking your tender green vegetables sous vide ensures that they retain their crispness and bright flavors. Try cooking up an extra-large batch, then keeping the vegetables chilled in the refrigerator for easy and healthy snacking, or to add to meals throughout the week.

4 cups fresh green beans, sugar snap peas, snow peas, asparagus, or fava beans
2 tablespoons olive oil
1 tablespoon lemon zest
2 tablespoons lemon juice
1 teaspoon salt
½ cup toasted almonds, roughly chopped

1 Preheat the water bath to 183°F.

2 Place the vegetables, oil, and lemon zest in a vacuum-seal bag and shake to coat. Seal the bag using a vacuum sealer on the moist setting.

3 Place the bag in the water bath and cook for 1 hour.

4 Remove the vegetables from the bag and place on a serving platter.

5 Drizzle with the lemon juice and sprinkle with the salt.

6 Top with the chopped almonds and serve.

Tip: *For many recipes, cooking for a little longer won't change the result, so you can be flexible with timing.*

Rugged Vegetables

YIELD: 6 TO 8 SERVINGS / ACTIVE TIME: 10 MINUTES / TOTAL TIME: 1 HOUR, 10 MINUTES

You can still achieve the delicious flavors of roasted vegetables without the danger of over- or undercooking them. These hearty root vegetables will retain more flavor and color than with traditional cooking methods, and you'll be amazed at how creamy and tender they come out.

1 pound root vegetables (potatoes, carrots, beets, parsnips, rutabagas, etc.)
2 tablespoons unsalted butter
½ teaspoon salt, plus additional as needed
Freshly ground black pepper

Tip: *Cook meals in batches to freeze or refrigerate. Reheat in the water bath.*

1 Preheat the water bath to 183°F.

2 Scrub or peel the vegetables and cut into bite-size pieces. Place the vegetables, butter, and ½ teaspoon of salt in a vacuum-seal or zip-top bag. Seal the bag with a vacuum sealer on the moist setting or using the displacement method.

3 Place the bag in the water bath and cook until fully tender, about 1 hour.

4 At this point, you can either finish the veggies immediately, or you can chill the bags in an ice bath for 30 minutes before storing them in the refrigerator for up to 4 days.

5 To serve, heat a large stainless steel or nonstick pan over medium-high heat and add the entire contents of the bag. Sauté until the veggies are brown, about 2 minutes.

6 Season with salt and pepper and serve.

Amazing Artichokes

YIELD: 4 SERVINGS / ACTIVE TIME: 10 MINUTES / TOTAL TIME: 40 MINUTES

As artichokes are one of my favorite foods, I've experimented with many ways of cooking them. I've learned that sous vide is truly the best way to prepare an artichoke. All the wonderful flavor that normally seeps out into the steaming water stays locked in the artichoke. It's like the most potent artichoke-y tasting artichoke you can imagine. Even better, the choke part comes away from the heart without a fight or need for a knife, leaving as much of the heart intact as possible.

2 globe artichokes
1 lemon, quartered
Salt and freshly ground black pepper
Hollandaise Sauce (page 208; optional)

1 Preheat the water bath to 183°F.

2 Peel the stems of the artichokes and remove the lower leaves. Cut the artichokes in half through the stem.

3 Place the artichokes in a vacuum-seal bag. Seal the bag using a vacuum sealer. Vacuum sealers work best for these, but you can use zip-top bags with the displacement method. Since artichokes like to float, add a weight to the bag, like a butter knife, before sealing, or a large binder clip on the outside.

4 Place the bag in the water bath and cook for 30 minutes.

5 Remove from the bag.

6 Serve with lemon, salt and pepper, and hollandaise (if using).

Asparagus with Egg and Crispy Prosciutto

YIELD: 4 SERVINGS / ACTIVE TIME: 15 MINUTES / TOTAL TIME: 40 MINUTES

This dish is a great way to experiment with "staged cooking," where you cook one part of the dish first, then lower the temperature of the water bath to finish cooking the rest. Since the lower temperature doesn't continue to cook the vegetables, it's a great way to keep everything warm and have it all ready at the same time.

1 bunch asparagus, trimmed and peeled
1 tablespoon butter
Salt
Ice
4 large eggs
2 teaspoons olive oil
4 slices prosciutto, cut into 1-inch strips horizontally (against the grain)
Freshly ground black pepper
Truffle oil (optional)

1 Preheat the water bath to 183°F.

2 Place the asparagus in a single layer in a large vacuum-seal or zip-top bag. Add the butter and sprinkle with a pinch salt. Seal the bag using a vacuum sealer or the displacement method.

3 Place the bag in the water bath and cook for 12 minutes.

4 Leaving the asparagus in the water bath, bring the temperature down to 167°F by adding ice, a little at a time. If you go too low, add warm water until you hit your mark.

5 Add the eggs and cook for 13 minutes.

6 Meanwhile, heat the olive oil in a skillet over medium-high heat until hot. Add the prosciutto and cook until crispy, 3 to 4 minutes. Transfer to a paper towel–lined plate.

7 When the eggs are done and the asparagus spears are tender, remove them all from the water bath. Remove the asparagus spears from the bag and arrange on a serving platter. Drizzle the butter from the bag over the spears. Season with salt.

8 Crack the eggs and slide them on top of the asparagus.

9 Top with crispy prosciutto and season with pepper and a drop of truffle oil (if using).

Tip: *To make removing the eggs from the water bath easier, you can place them in a zip-top bag and clip them to the side of water bath.*

Bánh Mì with Lemongrass and Barbecued Tofu

YIELD: 4 SERVINGS / ACTIVE TIME: 30 MINUTES / TOTAL TIME: 3 HOURS

Bánh mì can be served with almost any kind of meat, but the bread and the toppings are what make it iconic. Be sure to get great crusty baguettes and load up with more cilantro than you feel comfortable with; then you'll have something akin to what you'd get from a Vietnamese sandwich shop. This tofu recipe is super versatile, so don't feel limited to eating it only in sandwich form.

1 (14-ounce) package extra-firm tofu, drained and sliced into 4 planks
1 teaspoon olive oil

FOR THE LEMONGRASS SAUCE

¼ cup soy sauce
¼ cup olive oil
2 tablespoons sesame oil
2 lemongrass bulbs, minced
2 teaspoons garlic powder

Or use
⅔ cup barbecue sauce (page 212)

FOR THE DO CHUA (PICKLES)

½ cup sugar
1 teaspoon salt
2 cups water
1 cup white vinegar
2 cups julienned daikon (or cabbage)
2 cups julienned carrots

FOR ASSEMBLING

2 crusty baguettes, cut in half
¼ to ½ cup mayonnaise
1 medium cucumber, sliced lengthwise
1 or 2 jalapeño peppers, sliced
2 bunches of cilantro, leaves picked but left whole
Sriracha (optional)

1 Preheat the water bath to 180°F.

2 Line a sheet pan or cutting board with two layers of paper towels. Set the slices of tofu in a single row on top. Place another two layers of paper towels on top of the tofu. Set another sheet pan or cutting board on top and weigh it down with heavy books or cans. Press the tofu for about 30 minutes to extract extra moisture.

3 If using lemongrass sauce: In a small bowl, whisk together the soy sauce, olive oil, sesame oil, lemongrass, and garlic powder.

4 Remove the tofu from the sheet pan and place it in a single layer in a large vacuum-seal or zip-top bag. Add the lemongrass sauce or barbecue sauce. Seal the bag using a vacuum sealer or the displacement method.

5 Place the bag in the water bath and cook for 2 hours.

6 To prepare the pickles: In a large bowl, combine the sugar, salt, water, and vinegar. Add the daikon and carrots and toss. Cover and refrigerate for at least 1 hour. Drain completely before using.

7 Remove the tofu from the bag, reserving the cooking liquid.

8 Heat the olive oil in a stainless steel or cast iron skillet. Add the tofu and sear until crispy, about 30 seconds on each side. Add the reserved liquids and cook for 1 to 2 minutes.

9 To serve, slice the baguette halves lengthwise, leaving one side connected. Spread the mayonnaise on the top and bottom halves. Arrange the cucumber, do chua, tofu (with sauce), jalapeño, cilantro, and sriracha (if using), on one side of the baguette and close the other side.

Tip: *Preshredded coleslaw mix makes a great stand-in for the carrots and daikon.*

French Onion Soup

YIELD: 4 SERVINGS / ACTIVE TIME: 50 MINUTES / TOTAL TIME: UP TO 24 HOURS

There's a lot of debate around whether you can truly caramelize onions using sous vide. But by starting the process on the stove and finishing it in the water bath, you can achieve an amazing depth and complexity of flavor with very little effort. While technically the sous vide temperature is too low to caramelize, braising the onions before kick-starts the slow breakdown of the sugars and that sweet umami flavor. It's the perfect base for this French Onion Soup.

FOR THE CARAMELIZED ONIONS

4 tablespoons butter, divided
2 tablespoons olive oil
4 onions, thinly sliced
¼ teaspoon salt

FOR THE SOUP

1½ cups dry white wine (optional)
10 thyme sprigs
2 bay leaves
6 cups Beef Stock (page 206)
8 (½-inch) baguette slices
1 garlic clove, cut in half lengthwise
4 ounces Gruyère cheese, grated (about 1 cup)

1 Preheat the water bath to 186°F.

2 Heat a stainless steel or cast iron skillet over medium heat. Add 2 tablespoons of butter and the oil and heat until the butter melts. Add the onions and sprinkle with the salt.

3 Cook until the onions lose much of their liquid and become transparent. Let most of the liquid cook off, 5 to 7 minutes. Remove from the heat and cool completely.

4 Place the cooled onions in a vacuum-seal bag. Seal the bag using a vacuum sealer.

5 Place the bag in the water bath and cook for up to 24 hours. When the onions are done, you can remove them and place in the refrigerator for up to 3 days. Be sure to chill them in an ice bath first.

6 To make the soup, add the onions and their cooking liquids from the bag to a large stockpot over medium heat and bring to a boil. Add the wine (if using) and increase the heat to high. Cook until almost all the liquid has evaporated, 8 to 10 minutes.

7 Tie the thyme sprigs and bay leaves into a bundle with twine or in cheesecloth, or add to a tea strainer. Add the herb bundle and stock to the pot.

8 Bring to a simmer and cook, uncovered, until the stock is thickened, about 20 to 30 minutes.

9 Remove from the heat. Remove the herb bundle and whisk in the remaining 2 tablespoons of butter. Season with salt and pepper.

10 Heat the broiler.

11 Place the baguette slices on a baking sheet and toast under the broiler, about 1 minute per side. Rub one side of each toast with the garlic clove.

12 Place four ramekins or oven-safe bowls on a rimmed baking sheet and ladle in the soup. Top each serving with two garlic-rubbed toasts. Divide the cheese among the servings, covering the bread and some of the soup.

13 Carefully place the baking sheet back under the broiler until the cheese is melted and bubbling, 4 to 8 minutes. If you don't have ovenproof bowls, sprinkle the cheese on the toasts while they're still on the baking sheet and place back under the broiler for 2 minutes, then top each soup bowl with two slices of the toast.

Corn on the Cob Three Ways

YIELD: SERVES 4 / ACTIVE TIME: 10 MINUTES / TOTAL TIME: 40 MINUTES

Cooking corn sous vide means all the flavor stays within the kernels, making it juicier, sweeter, and more intense than any other corn you've had. If you have trouble with the bags floating, clip a heavy spoon to the bag with a large binder clip to help weigh it down.

MEXICAN STREET CORN

Mexican street corn is the king of the corn world. Sweet, salty, savory, and creamy, this dish hits all the notes. Though typically grilled, this version creates a much more tender, juicier corn on the cob. Slathered with rich mayo and sour cream, then topped with spices and cilantro, the corn is so delicious I know you're going to love it as much as I do.

¼ cup mayonnaise
¼ cup sour cream
½ teaspoon ancho or chipotle powder, plus more for serving
1 garlic clove, finely minced
¼ cup finely chopped cilantro
4 ears of corn, husks removed
½ cup finely crumbled cotija or feta cheese, plus more for serving
1 lime, cut into wedges

Tip: *If you're really missing that smoky grilled flavor, add a few drops of liquid smoke to the bags before cooking!*

1 Preheat the water bath to 183°F.

2 In a large bowl, stir together the mayonnaise, sour cream, chile powder, garlic, and cilantro.

3 Add the corn and mayo mixture to one or two vacuum-seal bags and roll the corn around to coat evenly. Seal the bags using a vacuum sealer.

4 Place the bag(s) in the water bath and cook for 30 minutes.

5 Remove the corn from the bag(s) and dust with cotija and extra chile powder. Serve with the lime wedges.

BASIC CORN ON THE COB

For those of you who aren't butter lovers, cooking the corn directly in its husk in the bag gives the kernels a more complex flavor with some nice grassy notes. Of course, you can still add butter afterward!

4 ears of corn, still in husks

1 Preheat the water bath to 183°F.

2 Trim the ends off the corn. Place the corn in one or two large vacuum-seal or zip-top bag(s). Seal the bag(s) using a vacuum sealer or the displacement method.

3 Place the bag(s) in the water bath and cook for 30 minutes.

4 Remove the corn from the bag(s), remove the husks, and serve.

EXTRA BUTTERY CORN ON THE COB

Okay, who are we kidding? Half the joy of corn on the cob is the butter! This recipe utilizes the benefits of vacuum sealing by getting the butter into every nook and cranny as it melts. You can even add your favorite herbs or seasonings to the bag for more flavor.

4 ears of corn, husks removed
2 tablespoons butter, plus more for serving
Salt

1 Preheat the water bath to 183°F.

2 Place the corn and butter in one or two large vacuum-seal or zip-top bag(s). Seal the bag(s) using a vacuum sealer or the displacement method.

3 Place the bag(s) in the water bath and cook for 30 minutes.

4 Remove the corn from the bag(s), sprinkle with salt, and serve with extra butter.

French Fries and Sweet Potato Fries

YIELD: 4 SERVINGS / ACTIVE TIME: 30 MINUTES / TOTAL TIME: 1 TO 1½ HOURS

Homemade fries, especially sweet potato fries, can be really hit or miss: Either they're too soggy or burnt to a crisp. By cooking them first at the temperature where the pectin starts to break down, and then following it up with a high-heat bake or deep fry, you can ensure a perfect flaky interior and a crispy crust. You've never had fries like this.

2 large russet potatoes or sweet potatoes, peeled and sliced into ¼-inch-thick sticks
2 quarts vegetable oil, for frying or 1 to 2 tablespoons for baking
Salt
Freshly ground black pepper
Your favorite seasoning blend (optional, to replace salt and pepper)

Tip: *I love the sweet potato fries tossed with smoked paprika or the Sweet and Spicy Rub (page 216).*

1 Preheat the water bath to 183°F.

2 Place the potatoes in a single layer in a large vacuum-seal or zip-top bag. Seal the bag using a vacuum sealer or the displacement method. Try to keep the fries in a single layer so that they don't overlap.

3 Place the bag in the water bath and cook for 45 minutes.

TO FRY

1 Prepare an ice bath. When the potatoes are done, remove the bag from the water bath and chill in an ice bath for 20 to 30 minutes while you prepare a pot for frying.

2 In a deep fryer or heavy pot, heat 2 quarts of oil to 350°F.

3 Remove the potatoes from the bag and pat dry.

4 Add a large handful of potatoes to the hot oil and cook until golden brown, 5 to 7 minutes. Remove the potatoes with a slotted spoon or spider, gently shaking off any excess oil, and let drain on a rack. Repeat until all of the potatoes are cooked.

5 Season with salt and pepper or your favorite seasoning blend.

TO FINISH IN THE OVEN

1 Preheat the oven to 450°F.

2 Lightly oil a baking sheet and place it in the oven until smoking hot.

3 When the potatoes are done cooking, remove the bag from the water bath. Remove the fries from the bag and gently pat dry with paper towels.

4 Carefully remove the cooking sheet from the oven, taking care with the hot oil.

5 Using tongs, gently coat the fries in the hot oil on the pan. Bake the fries for 5 minutes.

6 Carefully toss the fries again and cook for another 2 to 5 minutes, until crispy. Using a slotted spoon, transfer the fries to a paper towel–lined plate and immediately season with salt and pepper or your favorite seasoning blend.

Creamy Garlic Mashed Potatoes

YIELD: 4 SERVINGS / ACTIVE TIME: 15 MINUTES / TOTAL TIME: 1½ HOURS

There's nothing more comforting than digging into a heap of buttery, garlicky mashed potatoes. Avoid common mistakes that can result in lumpy or runny potatoes by cooking them right the first time! Add the potatoes and seasonings to the bag, and just walk away. Then just mash it all up together, and you're good to go. If you want to experiment with different flavors, this is the perfect way to try it!

4 large russet potatoes, scrubbed (or peeled if desired), and cut into ½-inch pieces
4 garlic cloves, peeled and smashed
6 tablespoons unsalted butter
1 cup whole milk
Salt
Freshly ground black pepper

Tip: *This recipe is also great with sweet potatoes! Just omit the milk and reduce the time to 45 minutes, or they'll get mushy. Add some maple syrup or brown sugar when you mash them if that's your jam.*

1. Preheat the water bath to 183°F.
2. Add the potatoes, garlic, butter, and milk to a zip-top bag and seal using the displacement method.
3. Cook for 1 hour (the potatoes can stay in the bath up to 2 hours).
4. Open the bag, strain the liquid into a small bowl, and reserve it.
5. Add the potatoes to a large bowl and roughly mash them with a fork or potato masher.
6. Gently stir the reserved liquid back into the mashed potatoes.
7. Season with salt and pepper.

Glazed Carrots

YIELD: 4 TO 6 SERVINGS / ACTIVE TIME: 10 MINUTES / TOTAL TIME: 1 HOUR, 10 MINUTES

Glazed carrots is a classic dish any time of year. But when you boil, steam, or sauté your carrots, you lose so much of their vibrant flavor. Now you get to keep all of their bright complexity and healthy nutrients intact while bumping up that sweetness profile with a brown sugar glaze that cooks right in the bag.

1 pound medium to large carrots, peeled and cut into ½-inch chunks (or use baby carrots)
2 tablespoons unsalted butter
1 tablespoon brown sugar (optional; this is what makes the glaze)
½ teaspoon salt, plus additional as needed
Freshly ground pepper
1 tablespoon chopped parsley (optional)

1 Preheat the water bath to 183°F.

2 Place the carrots, butter, sugar (if using), and ½ teaspoon salt in a single layer in one or two vacuum-seal bags. Seal using a vacuum sealer.

3 Place the bag(s) in the water bath and cook for 1 hour.

4 You can either finish the carrots immediately, or you can chill the bag(s) completely in an ice bath and store in the refrigerator for up to 4 days.

5 To serve, heat a large stainless steel or nonstick pan over medium-high heat and add the entire contents of the bag(s). Stir and heat until a glaze forms, about 2 minutes.

6 Season with salt and pepper, stir in the parsley (if using), then serve.

Carrot and Beet Salad

YIELD: 6 TO 8 SERVINGS / ACTIVE TIME: 20 MINUTES / TOTAL TIME: 1 HOUR, 20 MINUTES

In the same amount of time it will take you to make plain carrots, you can prepare this vibrant and flavorful salad. The reason for bagging the vegetables separately is because the bright red beet juice will stain the carrots. In order to keep the salad as colorful as possible, plate them together just before serving.

FOR THE CARROTS

1 pound carrots, peeled and cut into ½-inch chunks
2 tablespoons unsalted butter
1½-inch-thick orange slice
1 teaspoon chopped fresh tarragon
Salt
Freshly ground pepper

FOR THE BEETS

1 pound beets (3 to 4 large or 6 to 8 small), peeled and cut into ½-inch cubes
1½-inch-thick orange slice
½ teaspoon honey
½ teaspoon chopped fresh tarragon
Salt
Freshly ground pepper

FOR THE ASSEMBLY

½ cup crumbled ricotta salata

Tip: *This salad is also great served cold! After cooking, chill the carrots and beets in an ice bath for 30 minutes. Skip the pan-glazing step and toss together with a little extra orange juice and salt and pepper, and serve with ricotta.*

1 Preheat the water bath to 183°F.

2 Combine the carrots, butter, orange slice, and tarragon in a vacuum-seal bag. Sprinkle with salt and pepper. Seal using a vacuum sealer on the gentle setting.

3 Add the beets, orange slice, honey, and tarragon to a second vacuum-seal bag. Sprinkle with salt and pepper. Seal using a vacuum sealer on the gentle setting.

4 Place the bags in the water bath and cook for 1 hour.

5 Remove the bags from the water bath. Finish the salad immediately, or chill the bags in an ice bath and store in the refrigerator for up to 4 days.

6 When you're ready to finish the salad, add the contents of the bag with the carrots to a skillet over medium-high heat and simmer until the cooking liquid thickens and coats the carrots. Season with salt and pepper. Transfer to a platter and keep warm.

7 Repeat with the beets.

8 Add the beets to the carrots and top with the ricotta.

Perfect Roasted Potatoes

YIELD: 4 SERVINGS / ACTIVE TIME: 10 MINUTES / TOTAL TIME: 1 HOUR, 10 MINUTES

I love potatoes. Did you know they even have vitamins? I had to research it because I thought there's no way anything so tasty could be good for you. Plus, the cook time and temperature are perfect for a mélange of root vegetables. Toss in some parsnips, carrots, and onions for a side dish perfect for a Sunday roast.

4 large russet potatoes, cut into 2-inch pieces
4 fresh rosemary or thyme sprigs or favorite herbs
2 tablespoons fat (duck fat, butter, or olive oil)

1 Preheat the water bath to 183°F.

2 Add the potatoes, herbs, and fat in a single layer in a vacuum-seal bag. Seal using a vacuum sealer.

3 Place the bag in the water bath and cook for 1 hour.

4 When the potatoes are done cooking, heat a stainless steel or cast iron skillet over medium-high heat.

5 Open the bag and add the potatoes and fat to the hot skillet.

6 Shake the skillet around to smash up the potatoes, and cook until the edges are crispy.

Risotto with Porcini Mushrooms

YIELD: 4 SERVINGS / ACTIVE TIME: 30 MINUTES / TOTAL TIME: 1 HOUR, 15 MINUTES

Cooking risotto is traditionally a labor of love. It's all about standing over the stove and constantly stirring until your dish reaches the perfect consistency. Meanwhile, the rest of your menu is on standby! Using sous vide to cook your rice means that your time is free to prepare the rest of the meal. If you can't find dried porcini mushrooms, substitute dried pioppini, morels, black trumpets, or shiitake.

½ ounce dried porcini mushrooms
1½ tablespoons unsalted butter, divided
1½ teaspoons extra-virgin olive oil
½ cup chopped onion
8 ounces white mushrooms, thinly sliced
1 cup Arborio rice
½ cup dry white wine or vermouth
2 cups vegetable broth
⅓ cup grated Parmesan cheese, plus more as needed
¼ cup thinly sliced scallions, green and white parts

1 Preheat the water bath to 181°F.

2 Place the porcini mushrooms in a small bowl with ½ cup boiling water. Let soak for 30 minutes. Remove the mushrooms and roughly chop.

3 Strain the soaking water through a fine-mesh strainer into a bowl and reserve.

4 In a large stainless steel or nonstick skillet, heat ½ tablespoon of butter and the olive oil over medium-high heat. Add the onion and mushrooms and cook, stirring frequently, until the mushrooms have released their liquid and the liquid has evaporated, about 10 minutes.

5 Stir in the Arborio rice and cook for 1 minute. Add the white wine and reserved porcini liquid.

6 Bring the skillet to a rapid simmer and cook until the liquid is almost completely evaporated, about 10 minutes.

7 Stir in the chopped porcini and remove from the heat. Transfer the rice mixture to a large zip-top bag.

8 Add the broth and seal the bag using the displacement method.

9 Place the bag in the water bath and cook for 45 minutes.

10 Occasionally remove the bag from the bath and agitate it with your hands (squish it around a bit).

11 When the risotto is done cooking, remove the bag from the water bath.

12 Open the bag and sample the rice; if it is not quite tender, return it to the water bath for 5 minutes. Repeat until the rice is cooked. When the rice is tender, empty the entire contents of the bag into a large bowl.

13 Stir in the remaining 1 tablespoon of butter, the Parmesan cheese, and the scallions.

14 Serve topped with extra grated Parmesan cheese.

Note: *Dried porcini lack the rich, buttery texture of plump, freshly gathered ones. But it's difficult to find them fresh, so where flavor counts, as in this risotto, dried will do fine. Reconstituting them with water will bring out and intensify their earthy flavor.*

Slow-Braised Green Beans

YIELDS: 4 SERVINGS / ACTIVE TIME: 10 MINUTES / TOTAL TIME: 3 HOURS

Green beans are so simple and fresh, it wouldn't seem like cooking them sous vide could do anything to improve upon them. But when you seal in all of their juices, their bright, grassy flavor really shines through. And it's still as simple as anything—just bag, seal, and cook. But in my house, we're huge fans of long-cooked green beans. The long time spent braising in liquid makes them rich and tender, and cooking them sous vide means you're still able to capture the brightness of the beans.

1 pound green beans, trimmed
1 (16-ounce) can tomatoes, whole or crushed (I prefer San Marzano)
½ cup broth (beef, chicken, or vegetable)
1 onion, thinly sliced
3 garlic cloves, thinly sliced
½ teaspoon smoked paprika (optional)
½ teaspoon salt, plus additional as needed
Freshly ground black pepper

1 Preheat the water bath to 183°F.

2 Combine the green beans, tomatoes, broth, onion, garlic, paprika (if using), and ½ teaspoon of salt in a large zip-top bag. Seal the bag using the displacement method.

3 Place the bag in the water bath and cook for 3 hours.

4 When they're done, transfer the green beans and sauce to a large serving bowl. Season with salt and pepper.

Tip: *For traditional green beans, cook with 1 teaspoon olive oil for only 30 to 45 minutes. Make your own variations by including your favorite herbs, citrus, and seasonings!*

Ratatouille

YIELD: 4 TO 6 SERVINGS / ACTIVE TIME: 20 MINUTES / TOTAL TIME: 1 HOUR

I never thought I would love ratatouille, but honestly, it's one of my favorite ways to eat eggplant and zucchini. Go ahead and mix up the amounts of each vegetable to suit your taste. I like to serve this with just a bit of fresh goat cheese to perk up the flavors, but you could swap in Parmesan for a milder finish or leave cheese out altogether.

1 tablespoon olive oil
1 small onion, finely chopped
2 garlic cloves, sliced thinly
1 small eggplant, cut into ½-inch dice
1 yellow squash, cut into ½-inch dice
1 zucchini, cut into ½-inch dice
2 red bell peppers, cut into ½-inch dice
1 cup tomato purée
2 thyme sprigs
Salt
Freshly ground black pepper
Fresh goat cheese, for serving (optional)

Tip: *For added flavor, sear all of the vegetables for a few minutes before adding them to the bag.*

1 Preheat the water bath to 183°F.

2 In a stainless steel or nonstick skillet, heat the oil over medium-high heat. Add the onion and cook until soft and golden, 5 to 7 minutes. Add the garlic and cook until fragrant, about 1 minute.

3 Place the eggplant, squash, zucchini, bell peppers, tomato purée, thyme, and onion-garlic mixture in a vacuum-seal or zip-top bag. Seal using a vacuum sealer on the gentle setting or using the displacement method.

4 Place the bag in the water bath and cook for 40 minutes.

5 Remove the bag from the water bath and pour the veggies into a serving dish. Season with salt and pepper. Serve with a dollop of fresh goat cheese (if using).

7 DESSERTS

Sous vide cooking allows you to make the fanciest desserts of your life with minimal effort. Because of how precisely the water bath can treat eggs, tricky dishes like custards, puddings, and crème brûlée become foolproof. Prepare gorgeous poached fruits or perfect pie fillings, and top them off with your own sous vide ice cream.

Classic Crème Brûlée

YIELD: 6 SERVINGS / ACTIVE TIME: 20 MINUTES / TOTAL TIME: 1 HOUR, 20 MINUTES

Crème brûlée is a classic, luscious pudding with a caramelized crackle top, traditionally reserved for fancy French dining. The custard needs to be perfectly cooked without curdling and baked in a water bath in the oven until just set and not overdone. What dessert could be better suited to cooking sous vide? By maintaining a precise temperature in the water bath, you can turn out perfectly smooth, custardy crème brûlée every time. Now go get that kitchen torch!

4 egg yolks
¼ cup granulated sugar
2 cups heavy cream
1 teaspoon pure vanilla extract
2 tablespoons granulated sugar (for finishing)

1 Preheat the water bath to 176°F.

2 Place the egg yolks and sugar in a mixing bowl and whisk until the mixture is light and frothy and the sugar has dissolved.

3 Pour the cream into the egg mixture and whisk until combined.

4 Skim off any foam from the surface of the mixture. Stir in the vanilla.

5 Pour the mixture into six half-pint, wide-mouthed canning jars. Pour in a slow, steady stream to prevent bubbles from forming on the surface of the custard. If this happens, gently tap the jar on the counter to remove any bubbles.

6 Close the lids just tight enough for them to stay on, so any air will be able to escape from the submerged jars. If you close them too tightly, the trapped air will press against the glass and could crack or break your jars.

7 Carefully set the jars in the water bath and cook for 1 hour, until the custard has set. The centers should wiggle slightly when shaken but not be soupy.

8 Remove the jars from the bath and rest them at room temperature.

9 Prepare an ice bath. Once the jars are cool to the touch, transfer them to the ice bath to chill.

10 When the jars are cold, go ahead and tighten the lids. They will last up to a week, sealed, in the refrigerator.

11 When you're ready to serve, sprinkle the top of each custard with 1 teaspoon of sugar and caramelize it with a kitchen torch. If you don't have access to a torch, you can place the custards under a hot broiler.

Tip: *While traditional crème brûlée recipes call for scalding the cream and tempering the eggs, this step is necessary only if you'd like to infuse the cream with flavors like vanilla beans, whole spices, or coffee, or add chocolate. For this simplest crème brûlée around, no preheating is necessary.*

Chocolate Pots de Crème

YIELD: 6 SERVINGS / ACTIVE TIME: 20 MINUTES / TOTAL TIME: 5½ HOURS

There's nothing more luscious than dipping your spoon into a rich serving of pot de crème. Adding dark chocolate ups the decadence of this already glorious dessert that gets even more depth from just a hint of espresso powder. Serve with whipped cream and chocolate shavings for a deluxe treat.

6 large egg yolks
⅓ cup sugar
1 tablespoon water
½ teaspoon instant espresso powder
1 tablespoon vanilla extract
1 teaspoon salt
2 cups heavy cream
½ cup whole milk
5 ounces bittersweet chocolate, finely chopped

1 Preheat the water bath to 180°F.

2 In a blender or food processor, blend the egg yolks, sugar, water, espresso powder, vanilla, and salt until smooth, about 30 seconds. Transfer to a large bowl.

3 Combine the cream and milk in a small saucepan. Bring to a simmer over medium-high heat and then remove from the heat.

4 Place the chocolate in a medium bowl and place a fine-mesh sieve on top of the bowl.

5 Whisk the warm milk into the egg mixture very slowly to prevent the eggs from curdling.

6 Pour the custard through the fine-mesh sieve over the chocolate. Let stand for 5 minutes, then gently whisk until smooth. Let the custard cool for 20 minutes, letting bubbles rise to the surface.

7 Skim away any bubbles and divide the mixture among six jars, pouring in a slow, steady stream to prevent bubbles from forming on the surface of the custard. If this happens, gently tap the jars against the counter to remove any air bubbles.

8 Close the lids on the jars just tight enough for them to stay on, so any air will be able to escape from the submerged jars. If you close them too tightly, the trapped air will press against the glass and could crack or break your jars.

9 Carefully set the jars in the water bath and cook for 1 hour.

10 Remove the jars from the bath and let rest at room temperature for 10 minutes.

11 Prepare an ice bath. Transfer the jars to the ice bath to cool completely.

12 Chill for up to 4 hours before serving. Once the jars are cold they will last up to a week, sealed, in the refrigerator.

Individual Cherry Cheesecakes

YIELD: 4 SERVINGS / ACTIVE TIME: 40 MINUTES / TOTAL TIME: 2 HOURS

Nothing ends the night quite like your own individual cheesecake served up in a jar. There's something so comforting and quaint about holding your own little jar of sweetness with a slightly boozy topping. You can switch up the berries for your favorite fruit, and omit the alcohol. You might just want to make a double batch, because this will stay good in your refrigerator for up to a week—but I doubt they'll last that long!

FOR THE GRAHAM CRACKER CRUST

¼ cup crushed graham crackers or ginger snaps
2 tablespoons melted butter
½ tablespoon granulated sugar

FOR THE CHEESECAKE FILLING

12 ounces cream cheese
½ cup granulated sugar
¼ cup sour cream
2 eggs
1 tablespoon vanilla extract
Zest of one lemon, finely chopped
¼ cup lemon juice (about 2 lemons)

FOR THE TOPPINGS

2 cups frozen or fresh cherries, pitted
¼ cup granulated sugar
¼ cup brandy (optional)
1 tablespoon cornstarch
Whipped cream

1 Preheat the water bath to 176°F.

2 To make the crust, in a medium bowl, combine the graham crackers, butter, and sugar with a fork to mix thoroughly. Divide the crumbs evenly among four 8-ounce canning jars, and firmly press the crust into the bottom of each jar.

3 To make the filling, add the cream cheese, sugar, and sour cream to the bowl of an electric mixer. Beat until smooth. Add the eggs one at a time, beating well after each addition.

4 Beat in the vanilla, lemon zest, and lemon juice and mix until you have a smooth and creamy mixture, 5 to 7 minutes. Divide the mixture evenly among the jars, filling them up to the bottoms of the threads. Close the jars.

5 Carefully set the jars in the water bath and cook for 90 minutes.

6 Remove the jars from the bath and let rest until they reach room temperature before putting them into the refrigerator to cool further.

7 While the cheesecakes cool, make the topping. Add the cherries to a saucepan over medium heat. When the cherries begin to release their liquid, after 3 to 5 minutes, add the sugar and brandy (if using), and stir to combine.

8 Bring to a boil, stirring frequently. Reduce the heat and simmer until the cherries are tender, 5 to 7 minutes.

9 Add the cornstarch a little at a time, stirring constantly until the liquid has thickened to a syrupy consistency.

10 Remove from the heat and let cool.

11 To serve, remove the ring and lid from each jar. Spoon some cherries on top of the cheesecake. Top with whipped cream.

Coconut Rice Pudding with Cardamom & Crispy Coconut

YIELD: 4 SERVINGS / ACTIVE TIME: 15 MINUTES / TOTAL TIME: 2 TO 4 HOURS

Give this classic, cozy childhood treat an update with coconut milk and cardamom. Arborio rice yields the creamiest results, but for even more layers of flavor, substitute jasmine rice. Experiment with the spices, too—swap cinnamon for the cardamom, add in some saffron for extra color, or try maple syrup instead sugar. And being vegan and dairy-free, it makes a great dessert for dinner guests who might have dietary sensitivities.

2 (14-ounce) cans regular unsweetened coconut milk
½ cup sugar
1 cup unsweetened shredded coconut, divided
½ cup Arborio rice
½ teaspoon salt
1 teaspoon vanilla extract
8 cardamom pods, lightly crushed (optional)

Tip: *For a charming presentation, make this recipe in four individual 8-ounce canning jars.*

1 Preheat the water bath to 180°F.

2 In a large zip-top bag, combine the coconut milk, sugar, ½ cup of shredded coconut, rice, salt, vanilla, and cardamom (if using). Seal the bag using the displacement method.

3 Place the bag in the water bath and cook for 2 hours, agitating occasionally. After 2 hours, open the bag and taste the pudding. If the rice isn't fully cooked yet and needs more time, reseal and continue to cook, for up to 4 hours, checking intermittently for doneness.

4 While the rice pudding cooks, in a medium saucepan, toast the remaining ½ cup of coconut over medium heat, stirring constantly, until fragrant and golden, about 4 minutes. Transfer to a plate to cool.

5 Remove the cardamom pods and spoon the rice pudding into bowls, garnish with the toasted coconut, and serve warm or chilled.

Velvet Apricots with Rose Water and Pistachios

YIELD: 8 SERVINGS / ACTIVE TIME: 15 MINUTES / TOTAL TIME: 30 MINUTES

Apricots can be hit or miss when eaten fresh, which is why they're so often sold dried. And they can easily become overcooked and mushy when cooked using traditional methods. Holding the apricots at low temperatures using sous vide preserves all of the apricots' tenderness and vibrant color and highlights their bright flavor. Cook with a sweet and gently spiced syrup for a wonderful, simple dessert. Rose water complements the apricots nicely, but feel free to omit it if you can't find it.

8 ripe, firm apricots, halved with the stones removed
2 cups tap water
⅓ cup honey
¼ teaspoon ground cardamom
1 tablespoon rose water (optional)
¼ cup shelled raw pistachios, for garnish
Vanilla ice cream or whipped cream, to serve

1 Preheat the water bath to 167°F.

2 Place the apricot halves, tap water, honey, cardamom, and rose water (if using), in a single layer in a vacuum-seal or zip-top bag. Seal the bag using either a vacuum sealer on the gentle setting or the displacement method.

3 Place the bag in the water bath and cook for 18 to 20 minutes.

4 In a dry nonstick pan over medium-high heat, toast the pistachios, about 5 minutes.

5 When toasted, transfer the pistachios to a cutting board and roughly chop.

6 Remove the apricots from the bag, reserving the syrup, and serve alongside ice cream or whipped cream. Drizzle with the syrup from the bag and top with the toasted pistachios.

Lemon Curd Pavlova

YIELD: 6 SERVINGS / ACTIVE TIME: 40 MINUTES / TOTAL TIME: 4 HOURS

This lemon curd is as bright in flavor as it is in color, and you'll want to spread it on everything! If you've ever wondered what to do with leftover egg whites, this delicious pavlova—a deceptively simple, cloudlike meringue—is your answer. But if you're looking for a more hands-off dessert to whip up, you can also skip the pavlova and fill some premade mini pie shells or individual serving cups with the lemon curd and top with some fresh berries.

FOR THE LEMON CURD

⅔ cup granulated sugar
1 tablespoon cornstarch
⅛ teaspoon salt
⅓ cup freshly squeezed lemon juice
¼ cup (½ stick) unsalted butter
3 large egg yolks
2 teaspoons grated lemon zest

FOR THE PAVLOVA

1 cup granulated sugar
1 tablespoon cornstarch
3 large egg whites, at room temperature
Pinch salt
1 teaspoon freshly squeezed lemon juice

FOR SERVING

1 cup heavy whipping cream
1 pint fresh blackberries or blueberries

TO MAKE THE LEMON CURD

1 Preheat the water bath to 165°F.

2 To make the lemon curd, place all the ingredients into a zip-top bag. Seal using the displacement method.

3 Place the bag in the water bath and cook for 45 minutes to 1 hour.

4 Remove the lemon curd from the water bath, pour into a blender, and process until fully emulsified, about 30 seconds to 1 minute. The color of the liquid will lighten as you blend. Stop blending when the color stops changing.

5 Let the lemon curd cool completely, or chill in an ice bath, and then pour into a container or individual canning jars to set. Chill for up to 4 hours before serving.

6 Lemon curd will last up to 2 weeks in the refrigerator.

TO MAKE THE PAVLOVA

1 Preheat the oven to 300°F.

2 Place a sheet of parchment paper on a sheet pan. Draw a circle on the paper, using a dinner plate as a guide. Turn the paper over so the pencil marks are on the underside of the parchment.

3 To make the meringue, in a small bowl, whisk together the sugar and cornstarch.

4 In the bowl of a stand mixer fitted with a whisk attachment, beat the egg whites with a pinch salt on medium-high speed for about 3 minutes, or until small, soft peaks form.

5 Gradually add in the sugar mixture, one spoonful at a time, beating well after each addition.

6 Add the lemon juice and beat at high speed until the meringue is glossy and holds stiff peaks, about 5 minutes.

7 Spoon the mixture inside the circle drawn on the parchment paper. Using the circle as a guide, spread the mixture outward, keeping the sides higher than the center.

8 Bake until the meringue is a pale golden hue and has a crust, about 45 minutes.

9 Turn the oven off and let the meringue cool for 1 hour with the oven door propped open with a wooden spoon or folded towel.

TO SERVE

1 Beat the cream until it just holds stiff peaks, then fold ¼ cup of the beaten cream into the curd to lighten it up.

2 Spoon the curd into the meringue and mound the berries on top. Serve with the remaining whipped cream on the side.

Every Flavor Ice Cream

YIELD: 4 TO 6 SERVINGS / ACTIVE TIME: 20 MINUTES / TOTAL TIME: 6 HOURS

Sous vide is the easiest way to make creamy, luscious ice cream. Cooking at perfectly controlled temperatures means you'll never overheat the eggs or curdle your cream base. And since no one agrees on what's the best flavor (it's mint chip, obviously), I've provided a list of suggestions to try! Start with this base recipe and add your favorite mix-ins. You can even split the batch between bags and make two (or more) at once. I'm definitely screaming for this ice cream.

ICE CREAM BASE

1 cup whole milk

⅔ cup sugar

⅛ teaspoon fine sea salt

6 large egg yolks

2 cups heavy cream

Tip: *If you don't have an ice cream maker, flatten the bag and place in the freezer for 2 hours. Remove the bag and break up the ice cream base. Place the mixture into a food processor and process until creamy. Pour the creamy base into a container like a bread pan, cover with plastic wrap, and freeze. After 8 hours, you should have perfect, soft ice cream!*

1 Preheat the water bath to 185°F.

2 In a large bowl, whisk together the milk, sugar, salt, and yolks until thoroughly mixed. Whisk in the cream.

3 Pour the ice cream mixture into a zip-top bag. Seal using the displacement method.

4 Add the bag to the water bath and cook for 1 hour. Agitate the bag once or twice to ensure even cooking.

5 Prepare an ice bath. Remove the ice cream base from the water bath and chill in the ice bath until cold.

6 Remove the ice cream base from the bag and blend with a hand blender before churning it in an ice cream maker.

7 Freeze in the ice cream maker until firm.

VANILLA

To the base, add 1 vanilla pod, split and scraped. Add the seeds and pod to the bag. Strain before churning.

MINT CHOCOLATE CHIP

1 In a food processor, pulse together 1 cup of washed and dried fresh mint leaves with ⅔ cup of granulated sugar until pulverized and bright green. Add to the base.

2 Strain after cooking.

3 Stir in 8 ounces of chocolate chips and chill.

COFFEE

1 To the ice cream mixture, add ½ cup of coarsely ground coffee beans.

2 Strain before freezing.

GREEN TEA

1 To the ice cream mixture, add ¼ cup of loose-leaf green tea.

2 Strain before freezing.

CHOCOLATE

1 In a saucepan, bring ¾ cup of cream and 3 tablespoons of Dutch processed cocoa powder to a simmer.

2 Put 1 cup of chopped chocolate in a mixing bowl.

3 Pour the hot cocoa cream over the chocolate and stir until melted and smooth.

4 Make the ice cream base using these substitutes: 1½ cups of milk, ¾ cup of sugar, and no heavy cream.

5 Stir the chocolate mixture, ¾ cup of sour cream, and 1 teaspoon of vanilla extract into the base.

6 Strain before freezing.

PEANUT BUTTER

1 Make the ice cream base using 2 cups of milk and 1 cup of cream.

2 Whisk 1 cup of smooth peanut butter and ½ teaspoon of vanilla extract into the ice cream base.

Perfect Apple Pie

YIELD: 8 SERVINGS / ACTIVE TIME: 30 MINUTES / TOTAL TIME: 3 HOURS

The best baking apples offer a balance of sweet and tart flavors and a firm flesh that won't break down in the oven. I love Braeburn, Pink Lady, and Honeycrisp apples, but you can use your favorite, provided they're firm enough to hold up to baking. Use a blend of varieties for a richer flavor. Parcooking the apples sous vide will help them keep their shape, and the cornstarch and sugar blend creates an addictive gooey filling.

5 pounds apples, peeled, cored, and sliced ½ inch thick
½ cup sugar, plus 1 tablespoon
2 tablespoons cornstarch
½ teaspoon ground cinnamon
1 teaspoon grated lemon zest
2 teaspoons freshly squeezed lemon juice
Pie dough for a double-crust pie (store-bought or homemade)
1 egg white, lightly beaten

1 Preheat the water bath to 160°F.

2 In a large bowl, toss the apple slices with ½ cup of sugar, the cornstarch, the cinnamon, the lemon zest, and the lemon juice until well coated. Let rest for 10 minutes.

3 Transfer the apple mixture to a vacuum-seal bag. Seal using a vacuum sealer.

4 Place the bag in the water bath and cook for 1 hour.

5 Transfer the apple mixture and any liquids to a large Dutch oven and cook over medium-high heat, stirring frequently, until the juices thicken, about 10 minutes.

6 Transfer the apples to a rimmed baking sheet, spread out into a single layer, and allow to cool completely, about 1 hour.

7 Preheat the oven to 425°F.

8 Roll one pie dough disk into a 12-inch circle.

9 Lay the dough in a 9-inch pie plate. Add the pie filling.

10 Roll the remaining pie dough disk into a 12-inch circle and place over the apples. Fold the edges of both pie crusts down together, tucking them in between the bottom crust and the pie plate. Using your fingers or a fork, crimp the edges to seal.

11 Cut five slits in the top with a sharp knife.

12 Using a pastry brush, evenly coat the top of the pie with the lightly beaten egg white. Sprinkle the top with the remaining 1 tablespoon of sugar.

13 Set the pie on a sheet tray to make it easier to handle. Transfer to the oven and bake until light golden brown, about 20 minutes.

14 Reduce the heat to 375°F and continue baking until the crust is a deep golden brown, about 25 minutes longer. Remove the pie from the oven and allow to cool at room temperature for at least 4 hours before serving.

Poached Pears with White Wine and Ginger Sauce

YIELD: 4 TO 8 SERVINGS / ACTIVE TIME: 30 MINUTES / TOTAL TIME: 1 HOUR

There's something about the subtle sweetness and delicate texture of pears that makes them the perfect fruit for poaching. Whether you enrobe them in honey, wine, or port, you can completely transform the flavor, color, and texture of pears for a decadent dessert.

FOR THE POACHING LIQUID

2 cups dry white wine
1 cup sugar
2 strips lemon peel
2 tablespoons thin strips fresh ginger

FOR THE PEARS

4 pears, preferably Bosc or Bartlett, peeled
Mascarpone Topping (recipe follows; optional)

MASCARPONE TOPPING

1 (8-ounce) container mascarpone cheese, at room temperature
½ cup heavy cream
Pinch cinnamon
½ cup powdered sugar

1 Preheat the water bath to 185°F.

2 In a medium pot, combine the wine, sugar, lemon peel, and ginger and simmer until reduced by half, about 15 minutes.

3 Strain the cooking liquid.

4 Fill a container with water. Peel the pears and place in the water to keep them from browning.

5 If you like, using a clean sponge or kitchen towel, scrub the surface of each pear to give it a smoother appearance.

6 Place the pears and poaching liquid in a zip-top bag. Seal using the displacement method.

7 Add the bag to the water bath and cook the pears for 30 minutes, or until soft to the touch when squeezed.

8 Prepare an ice bath. When the pears are done cooking, transfer the bag to the ice bath and chill until cool.

9 Carefully remove the pears, reserving the juices. Set aside.

10 Transfer the juices to a pot and reduce on the stove for about 10 minutes, until they form a light syrup.

11 Cut the pears into thin slices and arrange on a plate. Pour the syrup over the pears and serve as is or top with the mascarpone topping or sous vide vanilla ice cream!

FOR THE MASCARPONE TOPPING

1 Whisk together the mascarpone cheese, heavy cream, cinnamon, and powdered sugar until smooth.

2 Dollop over the pears.

Recipe Idea: *For a tropical twist, try slices of pineapple infused with a coconut rum syrup!*

Glazed Donuts

YIELD: ABOUT 2 DOZEN DONUTS + DONUT HOLES / ACTIVE TIME: 2 HOURS / TOTAL TIME: 3½ HOURS

I can't think of what I love more than glazed donuts. Making them at home can be time consuming, but the results will astound you. Using sous vide takes out the trickiest part of the process—heating the eggs and milk to *just* the right temperature without scalding the milk or killing the yeast. Not a fan of glazed donuts? Top the donuts with chocolate glaze or fill them with lemon curd (page 172). However you serve them, you'll love these foolproof donuts.

¼ cup (½ stick) butter
¼ cup vegetable oil
2⅓ cups whole milk
3 large eggs, beaten
3 (¼-ounce) packets active dry yeast (6¾ teaspoons)
7½ cups bread flour
¾ cup granulated sugar
1 teaspoon salt
Nonstick spray, as needed
Canola oil for frying

FOR THE GLAZE

3 cups powdered sugar
½ teaspoon salt
2¼ teaspoons vanilla extract
⅓ to ½ cup milk

FOR THE CHOCOLATE GLAZE

2 cups powdered sugar
1 cup Dutch processed cocoa powder
½ teaspoon salt
⅓ to ½ cup milk

TO MAKE THE DONUTS

1 Preheat the water bath to 100°F.

2 Add the butter, oil, milk, and eggs to a zip-top bag. Seal using the displacement method.

3 Place in the water bath and cook for 10 minutes.

4 Open the bag, add the yeast, and agitate the ingredients. Cook for another 10 minutes.

5 Meanwhile, in a large bowl sift or whisk together the flour, sugar, and salt.

6 Remove the yeast mixture from the water bath and pour into the bowl of a stand mixer fitted with a dough hook.

7 Turn the mixer on to medium-low speed and add the flour mixture to the yeast mixture, one large spoonful at a time, until it's completely incorporated.

8 Turn the speed up to medium-high and mix until the dough is smooth and pulls away from the sides of the mixing bowl, forming a ball on the hook, about 10 to 20 minutes.

9 Spray your work surface and the inside of a large mixing bowl with nonstick spray. Shape the dough into a ball; place it in the mixing bowl, turning to coat it on all sides with the cooking spray. Cover with plastic wrap, and set it in the refrigerator to rise for 1 hour.

10 After the dough has risen, turn the dough out onto a floured surface; roll it around lightly to coat with flour.

11 Gently roll the dough to ½ inch thick with a floured rolling pin.

12 Cut the dough with a floured donut cutter.

13 Separate the donuts and holes, as they take different frying times (but are equally delicious). Save your scraps—they are great both to test your fry time and to make into Monkey Bread (page 183).

14 Line a sheet pan with parchment paper, spray the paper with nonstick spray, and transfer the donuts to the pan, leaving a few inches between them.

15 Cover the donuts loosely with a clean kitchen towel and let them rise until they double, 30 to 60 minutes. (If you want to make these donuts for breakfast, let the donuts rise in the refrigerator overnight.)

16 To fry the donuts, use a deep pan or deep fryer to heat the oil to 350°F. A thermometer makes this part fool-proof, and you can monitor the heat to make sure it stays in this prime frying range.

17 One at a time, fry the donuts for 15 seconds, then flip with a pair of chopsticks and cook until the bottom is dark, about 80 seconds. Flip and cook for another 60 to 80 seconds. Donut holes will take half as long to fry. With a larger pan, you can fry 2 to 3 donuts at a time.

18 Transfer the donuts to a rack over a sheet pan or to a paper towel–lined plate to drain.

19 Let the donuts cool for 15 minutes before glazing. If they're too warm, the glaze won't stick.

20 If necessary, rewhisk the glaze, then dip the tops of the donuts in the bowl and place on a rack to dry.

continued ›

Glazed Donuts continued

TO MAKE THE GLAZE

1 In a small bowl, whisk together the powdered sugar and salt.

2 In another bowl, stir the vanilla into the milk. Add the milk mixture to the powdered sugar, whisking constantly, until the glaze is smooth and of the consistency you like. The smaller amount of milk will make a thicker glaze; the larger amount will make a thinner glaze.

TO MAKE THE CHOCOLATE GLAZE

1 In a small bowl, whisk together the powdered sugar, cocoa powder, and salt.

2 Add the milk to the powdered sugar mixture, whisking constantly, until the glaze is smooth and of the consistency you like. The smaller amount of milk will make a thicker glaze; the larger amount will make a thinner glaze.

Monkey Bread

YIELD: 4 TO 6 SERVINGS / ACTIVE TIME: 25 MINUTES / TOTAL TIME: 1 HOUR, 20 MINUTES (PLUS ABOUT 3 HOURS IF MAKING DOUGH FROM SCRATCH)

Monkey bread takes me back to childhood when my mom would open a tin of refrigerated biscuit dough and roll pieces of them in cinnamon sugar and butter before baking. Use your leftover donut scraps or make up a new batch of dough for a sweet breakfast or treat any time of day.

¼ batch donut dough (see previous recipe, page 180) or remaining uncooked donut dough scraps
½ cup granulated sugar
1 teaspoon ground cinnamon
⅓ cup chopped walnuts or chocolate chips (optional)
Cooking spray
¼ cup (½ stick) butter
½ cup brown sugar

1 Preheat the oven to 350°F.

2 Make the donut dough according to the recipe instructions (page 180) through step 9. Divide the dough into four pieces and reserve three of the pieces for another use. If you're using donut scraps, form the dough scraps into a ball. Roll the dough out into a sheet about ½ inch thick, and sprinkle the granulated sugar, cinnamon, and walnuts or chocolate chips (if using), over the top.

3 Fold the dough into thirds (like a letter), then use a long knife to cut the dough into squares about an inch wide.

4 Coat the inside of a loaf pan or Bundt pan with the cooking spray, then add the dough pieces to the pan. Cover with plastic wrap and let the dough rise for 15 minutes at room temperature.

5 Meanwhile, in a small saucepan or microwave-safe bowl, melt the butter with the brown sugar.

6 Pour the sugar mixture over the donut scraps in the pan.

7 Bake for 35 minutes.

8 Let the bread cool in the pan for 10 minutes, then turn it out onto a plate.

9 Pull the bread apart and serve with strong coffee.

8

JUST FOR FUN

If all this wasn't way too much fun already, here are some of my favorite (extra-fun) things to do. When you're ready to take your new skills to the next level, try your hand at homemade yogurt, pickles, and sauces. Imbue vibrant flavors into oils, vinegar, and even butter. Experiment with the perfect cup of tea, mulled wine, or infused cocktail. Just have fun.

Sous Vide Cocktails

YIELD: 8 CUPS / ACTIVE TIME: 5 MINUTES / TOTAL TIME: 3 HOURS

There's one thing sous vide actually does incredibly fast: infusions. And what better to infuse than alcohol? For optimal flavor, definitely use the finest liquors and organic fruits. Since the cooking times on these are so short and all you need to do is add to some liquor in mason jars, I don't see why you wouldn't take a Saturday afternoon to just go nuts creating your own concoctions.

1 For each recipe, preheat the water bath to 130°F.

2 Add all the ingredients to a 16-ounce wide-mouth canning jar. Close the lid on the jar just tight enough for the lid to stay on so any air will be able to escape from the jars when submerged in water. If you close the lid too tightly, the trapped air will press against the glass and could crack or break your jar.

3 Set the jar in the water bath and infuse for at least 1 hour and up to 3 hours.

4 Before using, strain the liquor.

5 Serve the infusion over one large ice cube and garnish with a twist, or add it to your favorite cocktail.

Tip: *Remove just the zest from the fruit with a sharp vegetable peeler. If you get too much pith (the inner white stuff), use a paring knife to remove it. Otherwise, the result will be too bitter.*

OLD FASHIONED IN A JAR

14 ounces bourbon

2 ounces (about 4½ tablespoons) sugar

Zest of 1 orange

6 Morello or maraschino cherries

BOURBON BACON COCKTAIL

(great as a base for Old Fashioneds or Bloody Marys!)

14 ounces bourbon

2 ounces (about 4½ tablespoons) brown sugar or maple syrup

Zest of 1 orange

1 strip cooked bacon

QUICK LIMONCELLO

14 ounces vodka

2 ounces (about 4½ tablespoons) sugar

Zest of 3 lemons

ORANGE CREAMSICLE

14 ounces vodka

4 ounces (about 9 tablespoons) sugar

1 vanilla bean, split and scraped

Zest of 3 oranges

JALAPEÑO TEQUILA

(makes amazing watermelon margaritas)

14 ounces tequila

1 jalapeño, sliced

Homemade Yogurt

YIELD: 8 CUPS / ACTIVE TIME: 15 MINUTES / TOTAL TIME: 6 TO 24 HOURS, DEPENDING ON DESIRED TASTE AND TEXTURE

I consume yogurt nearly every day, but making it from scratch always seemed a little daunting to me. Playing with active cultures in a low-temperature environment seems risky, but the precision of sous vide makes it extremely accurate and dead easy. The resulting yogurt is very much like store-bought yogurt: thick and creamy with no need to strain it.

8 cups whole milk
4 tablespoons plain yogurt with live, active cultures

1 Preheat the water bath to 110°F.

2 In a heavy-bottomed pot over medium heat, bring the milk to 180°F.

3 Stir regularly to prevent scorching, but if the milk does stick to the bottom of the pan, just leave it. Don't try to scrape it up.

4 Once the milk has reached temperature, allow it to cool to 110°F. (Place it in an ice bath if you want to speed it up.)

5 When the milk has cooled, add the yogurt to the pot and whisk thoroughly to combine.

6 Pour the mixture into clean, sterilized jars (4 pint jars or 8 half-pint jars). Close the lids on the jars just tight enough for the lids to stay on so any air will be able to escape from the jars when submerged in water. If you close them too tightly, the trapped air will press against the glass and could crack or break your jars.

7 Place the jars in the water bath and cook for 6 to 24 hours. The longer the yogurt sits, the thicker and more tart it becomes.

8 When done, remove the jars from the water bath and place in the refrigerator for at least 6 hours to halt culturing and set the yogurt. Store in the refrigerator for up to 2 weeks.

Tip: *For extra-thick Greek-style yogurt, place a fine-mesh strainer over a bowl and line it with two layers of cheesecloth. Spoon the yogurt into the lined strainer and allow it to drain for 2 hours or until the desired thickness is achieved. Use the whey that's drained from the yogurt in baked goods or smoothies!*

Overnight Oatmeal with Warm Berry Compote

YIELD: 2 SERVINGS / ACTIVE TIME: 10 MINUTES / TOTAL TIME: 6 TO 10 HOURS

Imagine waking up to a steaming bowl of slow-cooked oatmeal ready and waiting for you with no more effort than it takes to pour a bowl of cold cereal. While you dream away, your oatmeal is slowly cooking in your sous vide bath, ready to be devoured upon your awakening. The best part is you can make enough for your whole household and customize each serving.

FOR THE OATMEAL

1 cup rolled oats
3 cups water
1 pinch salt
1 pinch cinnamon
1 pinch nutmeg

FOR THE BERRY COMPOTE

1 cup frozen berries, thawed
2 tablespoons honey
Zest and juice of ½ lemon

Tip: *Switch up the berry compote with apples and cinnamon or frozen mango and pineapple.*

1 Preheat the water bath to 155°F.

2 Pour the rolled oats, water, salt, cinnamon, and nutmeg into a zip-top bag. Seal using the displacement method.

3 Pour all the berry compote ingredients into a separate zip-top bag. Seal using the displacement method.

4 Submerge the bags in the water bath and cook for 6 hours (the oatmeal can stay in the water bath up to 10 hours).

5 Agitate the oatmeal pouch and pour the oatmeal into two bowls.

6 Open the pouch with the stewed fruit, spoon the fruit compote on top of the oatmeal, and enjoy.

Pickles

YIELD: 1 PINT EACH / ACTIVE TIME: 15 MINUTES / TOTAL TIME: 2 HOURS, 45 MINUTES

Preparing pickles sous vide cuts down the cook time in a major way, and heating them at such low temperatures preserves their crispness and intense flavor. I simplified these recipes by keeping them each uniform with a savory or sweet brine formula, but you can experiment to find what you like best. The best part is that these are done in 2 hours and shelf-stable for up to 6 weeks!

1 For each recipe, preheat the water bath to 140°F.

2 Mix together the liquid ingredients (water and vinegar), sugar, and salt in a small bowl to make a brine.

3 Peel or cut away the rinds, stems, or any other parts of the vegetables that you don't want to eat. Rinse the fresh herbs.

4 Transfer the fruits and vegetables, along with any seasonings, to a sterilized 1-pint canning jar, taking care not to overfill it. The food and seasonings should be able to float around freely.

5 Leave about 1 inch of headspace in the jar.

6 Add the brine, stopping when the liquid level is about ½ inch from the top of the jar.

7 Close the lid on the jar just tight enough for the lid to stay on so any air will be able to escape from the jar when you submerge it in water. If you close it too tightly, the trapped air will press against the glass and could crack or break your jar.

8 Set the jar in the water bath and cook for 2½ hours.

9 Remove the jar from the bath and let rest at room temperature overnight.

SWEET PICKLES

½ cup white wine vinegar
½ cup water
¼ cup sugar
1 cucumber, peeled, seeded, and sliced

PICKLED JALAPEÑOS

½ cup vinegar
½ cup water
1 tablespoon sugar
½ teaspoon salt
6 jalapeños, sliced into ¼-inch slices
½ white onion, thinly sliced

DILL PICKLES

½ cup white wine vinegar
½ cup water
1 tablespoon sugar
½ teaspoon salt
1 small bunch of dill
1 cucumber, peeled, seeded, and sliced

ZESTY DILL PICKLES

½ cup white wine vinegar
½ cup water
1 tablespoon sugar
½ teaspoon salt
1 small bunch of dill
1 cucumber, peeled, seeded, and sliced
½ red onion, cut into ¼-inch slices
2 garlic cloves, thinly sliced
1 teaspoon black peppercorns
1 teaspoon dried red pepper flakes

PICKLED GRAPES

½ cup red wine vinegar
½ cup water
¼ cup sugar
½ teaspoon salt
½ teaspoon ground cinnamon
½ teaspoon ground coriander
1½ cups red seedless grapes
1 rosemary sprig

PICKLED RADISHES

½ cup vinegar
½ cup water
1 tablespoon sugar
½ teaspoon salt
8 ounces radishes, trimmed and sliced into ¼-inch slices

GIARDINIERA

(makes a double batch)

1 cup vinegar
1 cup water
¼ cup sugar
1 teaspoon salt
1 tablespoon whole black peppercorns
3 serrano or jalapeño peppers, sliced
1 green bell pepper, diced
1 red bell pepper, diced
1 carrot, diced
1 celery rib, diced
1 yellow onion, diced
½ cauliflower head, cut into florets
3 garlic cloves, minced
½ cup stuffed pimiento green olives, chopped
2½ teaspoons dried oregano
½ teaspoon celery seed
½ teaspoon dried red pepper flakes

Preserves and Chutneys

YIELD: 8 SERVINGS EACH / ACTIVE TIME: 15 MINUTES / TOTAL TIME: 2 HOURS, 15 MINUTES

The fun thing about preserves and chutneys is that they all cook using the same temperatures and techniques, so you can experiment with more than one at a time! Dried fruits lend themselves beautifully to making simple preserves. Their flavors are concentrated, and they don't require pectin to set like traditional jams. Plus, you don't have to wait until your favorite fruits are in season to make them. Chutneys are a savory medley of fruits and spices that are perfect for pork, duck, and wild game. They also pair beautifully with cheese and wine. For optimum flavor, toast the spices in a hot pan before adding to the mix.

1 For each recipe, preheat the water bath to 183°F.

2 Combine all the ingredients in a large vacuum-seal or zip-top bag. Seal using a vacuum sealer or the displacement method.

3 Place in the water bath and cook for 2 hours.

4 When done, remove the bag from the bath and let cool.

5 Optional: Add everything from the bag into a bowl and use a potato masher to break down the larger bits. For a more uniform purée, use a food processor.

6 Store in canning jars in the refrigerator for up to 3 weeks.

BRANDIED CHERRY PRESERVES

1 cup sugar

1 cup water

½ cup brandy

1 rosemary sprig

2 tablespoons butter

2 cups dried cherries

BALSAMIC-FIG PRESERVES

2 cups dried mission figs, chopped

1 cup water

1 cup sugar

½ cup balsamic vinegar

MANGO CHUTNEY

3 mangoes, peeled, pitted, and diced

1 teaspoon coriander

1 teaspoon cumin

½ teaspoon cinnamon

½ teaspoon cayenne pepper

½ teaspoon turmeric

¼ teaspoon cardamom

1 cup apple cider vinegar

1 cup brown sugar

2 garlic cloves, minced

1½ teaspoons minced ginger

APRICOT PRESERVES

2 cups dried apricots, chopped

1½ cups sugar

1 cup water

Zest and juice of 1 lemon

APPLE CHUTNEY

2 apples, peeled and diced

¼ cup orange juice

¼ cup brown sugar

1 tablespoon thyme leaves

1 tablespoon apple cider vinegar

1 tablespoon butter

Salt

Freshly ground black pepper

HOT PEPPER CHUTNEY

5 medium-size hot chili peppers, charred, peeled, and chopped

1 jar roasted red peppers, drained

1 medium red onion, chopped

½ cup brown sugar

1 tablespoon balsamic vinegar

1 rosemary sprig

½ teaspoon ground cinnamon

¼ teaspoon salt

¼ teaspoon freshly ground black pepper

Infused Oils and Vinegars

YIELD: 1 PINT EACH / ACTIVE TIME: 5 MINUTES / TOTAL TIME: 2 TO 3 HOURS

Infused oil and vinegar recipes add a tasty variety to salads, marinades, vegetables, and sauces, and since you control all the ingredients, you cook healthier meals. Different flavors are extracted more quickly from the different infusions at different temperatures, typically 131°F to 160°F for vinegars and 149°F to 176°F for oils. To simplify things, I opted for 150°F so you can try all infusions at once, but experiment with different temps to find your favorite. For the oils, use extra-virgin olive oil or avocado oil for optimal flavor.

1 For each recipe, preheat the water bath to 150°F.

2 Combine the ingredients in a gallon zip-top bag, or use sterilized pint-size canning jars. Seal using the displacement method, or screw the lids on lightly.

3 Submerge the bag or jars in the water bath and cook for 2 to 3 hours, depending on how strong an infusion you want.

4 Partway through the cooking process, squeeze the pouch or shake the jars to agitate the contents.

5 Strain the oil or vinegar through cheesecloth or a fine-mesh strainer and discard the solids.

6 Pour the infusion into a clean bottle, cap tightly, label, date, and store in the refrigerator for up to 6 weeks.

Tip: *Why no recipes for garlic-infused oil? The bacteria spores that cause botulism can spread in certain foods when not exposed to oxygen—as is the case when infusing garlic in oil. When homemade garlic-infused oil is left unrefrigerated or kept for too long, there is a chance of this bacteria growing. If you want to make your own infused garlic oil, you should prepare it fresh and use it right away. If you are saving any leftover garlic oil, refrigerate it right away and use within a week. But instead of garlic oil, why not make Garlic Confit (page 196)?*

HERB OIL

2 cups oil
6 fresh oregano or thyme sprigs, or 3 fresh rosemary sprigs

LEMON OIL

2 cups oil
Zest of 4 medium-size organic lemons

CHILI OIL

2 cups oil
1 tablespoon dried crushed red pepper flakes

BASIL OIL

2 cups oil
4 cups packed fresh organic basil

HERB VINEGAR

2 cups distilled white vinegar
3 oregano sprigs
3 rosemary sprigs
3 marjoram sprigs

BLACKBERRY-BASIL VINEGAR

2 cups white balsamic vinegar
1½ cups fresh blackberries
¼ cup basil leaves

RAISIN VINEGAR

2 cups apple cider vinegar
¾ cup organic raisins

ORANGE-ROSEMARY VINEGAR

2 cups white balsamic vinegar
Zest of 3 organic navel oranges
5 rosemary sprigs

RASPBERRY-MINT VINEGAR

2 cups white wine vinegar
1½ cups fresh raspberries
¼ cup fresh mint leaves

Garlic Confit

YIELD: 8 OUNCES / ACTIVE TIME: 10 MINUTES / TOTAL TIME: 4 HOURS

Garlic confit is, at its heart, extremely simple and extremely delicious. Whole garlic cloves get roasted in oil and then used on anything from toast to roast chicken. You can use an entire head of garlic with the top lopped off, or you can separate and peel the garlic cloves depending on what you prefer. I prefer to peel them so I can really stuff as many cloves as will fit into the jar, but it does take more effort.

1 cup peeled garlic cloves
¼ cup olive oil
Salt (optional)

Pro tip: *Try using avocado oil or duck fat instead of olive oil. Fresh herbs like thyme or rosemary make a flavorful addition to both the garlic and the oil.*

1. Preheat the water bath to 190°F.
2. Add the peeled garlic cloves and oil to a canning jar or zip-top bag. (Using a canning jar makes for easier storage.) Seal the jar with the lid or use the displacement method to seal the bag.
3. Submerge the jar or bag in the water bath and cook for 4 hours.
4. Prepare an ice bath.
5. Remove the canning jar or bag from the water bath and place in the ice bath for at least 15 minutes.
6. Season the confit with salt (if using). Store in the refrigerator for up to 2 weeks.

Mulled Wine Two Ways

YIELD: 4 SERVINGS / ACTIVE TIME: 15 MINUTES / TOTAL TIME: UP TO 3 HOURS, 15 MINUTES

Mulled wine is a holiday tradition, flavored with warm spices and fruits. Typically made with red wine and served warm, it is also delicious made using white wine and served chilled. Heating sous vide to infuse the wine prevents any flavor loss that normally occurs when the wine is brought to a boil, resulting in a fuller, more nuanced mulled wine. The only drawback is that your house won't be filled with the wonderful smell while it cooks.

- 1 bottle semidry red wine (try a semidry Cabernet Sauvignon) or rich white wine (try a Chardonnay)
- 2 cups apple cider
- ⅓ cup honey
- ⅓ cup fresh whole cranberries
- 1 orange, sliced, plus more for garnish
- 1 apple, sliced
- 4 cinnamon sticks, plus more for garnish
- 4 whole cloves
- 2 star anise pods
- 1 vanilla bean, split in half
- 1 tablespoon whole cardamom pods
- 1 teaspoon whole cloves
- 1 small knob of ginger

1. Preheat the water bath to 140°F.
2. Combine all the ingredients in a gallon-size zip-top bag. Seal using the displacement method.
3. Place the bag in the water bath and heat for up to 3 hours.
4. If you're serving it warm, strain the wine and pour into mugs. Garnish with cinnamon sticks and orange slices.
5. If you're serving the wine chilled, prepare an ice bath and chill the bag in the ice bath for 20 minutes. Strain the wine and pour into glasses over ice. Garnish with orange slices.

Precision Tea

YIELD: 8 OUNCES EACH / ACTIVE TIME: 5 MINUTES / TOTAL TIME: 30 MINUTES

Tea is a delicate leaf, and each strain has its preferred ideal steeping time and temperature. When you use boiling water, green teas can turn bitter and astringent. Even some black teas should be steeped in cooler than boiling water. If you don't have one of those fancy electric kettles you can use to dial in a specific temperature, you can put your sous vide machine to work for you.

Filtered water
2 grams tea leaves per 1 to 3 servings
8-ounce canning jars
Tea infuser

Note: *If you have quality tap water and a very clean sous vide setup, you can just heat the water in your bath to the appropriate temperature and use that to heat your tea. I recommend the canning jars for a more controlled environment.*

1 Fill the water bath with just enough water to come below the ring of the canning jar lids; otherwise the jars won't stand upright in the bath (not a huge deal, just irritating). Preheat the bath according to the proper temperature for your tea.

2 Fill the canning jars with filtered water and close the lids.

3 Place the jars in the water bath and heat for 10 to 15 minutes.

4 Put the tea leaves in your infuser.

5 Add the infuser to the heated jar of water and steep according to the ideal time for your tea. Transfer the infuser to your next jar for another cup. The leaves should be good for two or three steepings before deteriorating in flavor.

GYOKURO	122°F
GREEN TEAS	158 to 180°F
OOLONG, DARJEELING	194°F
OTHER BLACK TEAS	212°F

Coffee and Tea Butters

YIELD: 2 CUPS / ACTIVE TIME: 5 MINUTES / TOTAL TIME: 3 HOURS

Give your mornings an extra boost with caffeinated coffee- or tea-infused butters. Most compound butters blend chopped ingredients with cold butter, but using sous vide actually extracts the flavors from the add-ins to create perfectly smooth, creamy butter that's great in just about everything. Use the freshest coffee beans and best-quality butter you can find. This also works with nut butters and even coconut oil if you're looking for a dairy-free option.

1 Preheat the water bath to 194°F.

2 Pack the beans or leaves and butter (and honey for tea butter), together in a vacuum-seal or zip-top bag. Seal using a vacuum sealer or the displacement method. (Coffee beans are buoyant, so you may need to add a weight to the bag to help it sink. Add a butter knife to the inside of the bag or a heavy binder clip to the outside.)

3 Place the bag in the water bath and cook for 3 hours.

4 Strain the infused butter, agitating the beans or leaves to extract as much of the infused butter as possible.

5 Discard the beans or leaves.

6 Pour the butter into a canning jar and store in the refrigerator.

Tip: *This butter is great on steak, as a base to a buttercream frosting, slathered on toast, or with sous vide dinner rolls (which happen to cook at the same temp). Try the coffee butter in your next chocolate cake or brownie recipe in place of butter to subtly deepen the chocolate flavor.*

COFFEE BUTTER

1 cup whole coffee beans

2 cups (4 sticks) salted butter or creamy almond butter

TEA BUTTER

1 cup whole tea leaves (Earl Grey recommended)

2 cups (4 sticks) unsalted butter or creamy almond butter

¼ cup honey

Dinner Rolls

YIELD: 12 ROLLS / ACTIVE TIME: 45 MINUTES / TOTAL TIME: 7 HOURS

Although dinner rolls are simple enough to make on their own, cooking them sous vide makes the task even simpler, taking out all the guesswork of handling the yeast with the utmost care and turning out amazingly tender, moist bread every time.

2 large eggs, beaten
¾ cup milk
¼ cup (½ stick) butter, plus more for serving
2¼ teaspoons (1 packet) active dry yeast
3 cups bread flour
1 teaspoon salt
¼ cup sugar
Nonstick cooking spray

1 Preheat the water bath to 100°F.

2 Add the eggs, milk, salt, sugar, and butter to a zip-top bag. Seal using the displacement method.

3 Place the bag in the water bath and cook for 10 minutes.

4 Open the bag, add the yeast, and agitate the ingredients. Cook for another 10 minutes.

5 In the bowl of a stand mixer fitted with a dough hook, add the contents of the bag and half of the flour. Beat for 3 minutes on medium speed.

6 Gradually add the remaining flour and knead with the dough hook until smooth and elastic, 5 to 7 minutes more.

7 Transfer the dough to a lightly greased bowl. Cover with plastic wrap and let the dough rise in a warm place until doubled in size, 1½ to 2 hours.

8 Increase the temperature of the water bath to 195°F.

9 Generously grease 12 half-pint canning jars with nonstick cooking spray.

10 Turn the dough out onto a lightly floured work space and roll into a short log. Slice the log into 12 equal pieces. Gently knead each piece and roll into a ball. Transfer one dough ball to each of the prepared jars.

11 Cover the jars with a clean kitchen towel and let the dough rise until doubled in size, 45 minutes to 1 hour.

12 Close the lids on the jars just tight enough for the lid to stay on so any air will be able to escape from the jars when you submerge them in water. If you close them too tightly, the trapped air will press against the glass and could crack or break your jars.

13 Place the jars in the water bath and cook for 3 hours.

14 Remove the jars from the water bath and let cool.

15 Carefully remove the lids. Flip the jars and gently tap on a kitchen towel to remove the rolls. Serve with butter.

Tempering Chocolate

YIELD: 8 TO 16 OUNCES / ACTIVE TIME: 25 MINUTES / TOTAL TIME: 25 MINUTES

When you melt chocolate to change its shape or use it in a recipe, you are taking it out of temper. The heat causes the fat molecules to get jumpy, and if they aren't realigned correctly, you get what's called "bloom." But bloomed chocolate can be brought back into temper. It takes minimal effort to get there, and once your chocolate is melted sous vide and is in temper, it can hang out in the water bath indefinitely. Temper as much or as little as you need; I find that 8 to 16 ounces is plenty for most recipes.

8 to 16 ounces milk chocolate or dark chocolate, roughly chopped

1 Preheat the water bath to 115°F.

2 Put the chocolate in a vacuum-seal bag (chopping is optional but speeds up the melting process). Seal using a vacuum sealer (vacuum sealing works best here because any amount of water coming into contact with your chocolate will make it seize and ruin it).

3 Add the bag to the water bath and cook until completely melted, about 5 minutes.

4 Lower the sous vide cooker temperature to 81°F (add ice to the water bath until the correct temperature is reached), then increase the temperature to 90°F for dark chocolate or 88°F for milk chocolate and let the chocolate heat up.

5 Take the bag out of the water once every minute or two and squeeze it around to agitate it as it warms up. This will prevent crystals from forming. Do this two or three times while the chocolate heats and once more right before you use it.

6 Hold the chocolate at this temperature until you're ready to use it. Snip off the corner of the bag and pipe or drizzle the chocolate as desired. When you're done, all you have to do is squeeze the chocolate away from the open corner of the vacuum bag, stick it back in your vacuum sealer, and reseal the edge. No cleanup!

7 Drizzle melted chocolate on fresh strawberries or pastries, or pipe it into molds. The chocolate will set with a gorgeous shine and snap.

9

STOCKS, SAUCES, BROTHS, AND SPICE RUBS

Homemade stocks and broths are a kitchen staple and one of the easiest and most affordable ways to up the level of flavors in your dishes. Cooking broths and stocks sous vide allows the meats and vegetables to impart deep, rich flavors into the liquids without boiling or simmering, so they maintain maximum nutrients. While spice rubs are (obviously!) not cooked sous vide, they're a great tool to have on hand for bringing out flavor in your sous vide meats and vegetables. The beauty of making your own rather than buying commercial blends is that you can customize them exactly as you wish—say you don't like cumin or coriander or you love extra cayenne; it's no problem. You're in charge.

Beef Stock / Bone Broth

YIELD: 4 QUARTS / ACTIVE TIME: 30 MINUTES / TOTAL TIME: 26 HOURS

This recipe combines roasting with sous vide to bring out the maximum flavor and health benefits of the bones. Some sources suggest adding a couple of tablespoons of apple cider vinegar to help draw the minerals out of the bones and into your broth. Once chilled, this broth is super gelatinous and jam-packed with flavor. Have your butcher cut a large (3-pound) knuckle into 4 pieces to give the most exposure to the marrow as possible.

2½ pounds ground beef
3 pounds beef knuckle
2 pounds beef bones
⅓ cup tomato paste
4 medium onions, unpeeled and cut in half
1 whole head of garlic, separated into individual cloves, unpeeled
2 large carrots, roughly chopped
3 thyme sprigs
2 fresh bay leaves
1 teaspoon black peppercorns
3 quarts (12 cups) water or reserved liquids from slow-cooked meats
Salt
Freshly ground black pepper

1 Preheat the oven to 425°F.

2 To one end of a roasting pan, add the ground beef.

3 Slather the beef knuckle and beef bones in tomato paste, and place at the other end of the roasting pan.

4 Place the onions and whole garlic cloves among the beef bones.

5 Put the pan in the oven, and roast until the meat is dark brown, 1½ to 2 hours, turning the ground beef and the bones once or twice.

6 Preheat the water bath to 194°F.

7 Remove the pan from the oven and strain the fat from the ground beef.

8 Add the roasted meat, bones, garlic, and onion, plus the carrots, thyme, bay leaves, peppercorns, and water to a 2-gallon zip-top bag.

9 Pour a little tap water into the hot roasting pan and scrape up any caramelized bits. Add them to the bag. Seal the bag using the displacement method.

10 Place the bag in the water bath and cook for 24 hours.

11 Remove the bones using tongs and discard. Strain the stock, discarding the solids. Season with salt and pepper. If not using immediately, chill the stock in an ice bath for 30 minutes, then transfer to the refrigerator. Once chilled, the fat will be easier to skim from the top of the stock. Store in the refrigerator or freezer.

Chicken Stock

YIELD: 4 QUARTS / ACTIVE TIME: 30 MINUTES / TOTAL TIME: 14 HOURS

Chicken stock is cooked much like the beef stock. Substitute up to 10 pounds of chicken bones, necks, wings, and/or legs and omit the tomato paste. Reduce the time in the water bath to 12 hours and otherwise follow the same recipe. How easy is that?

8 to 10 pounds chicken parts: bones, necks, wings, or legs
4 medium onions, cut in half
1 whole head of garlic, separated into individual cloves, unpeeled
2 large carrots, roughly chopped
3 thyme sprigs
2 fresh bay leaves
½ teaspoon black peppercorns
3 quarts (12 cups) water
Salt
Freshly ground black pepper

1 Preheat the oven to 425°F.

2 To a roasting pan, add the chicken parts, unpeeled onion halves, and whole garlic cloves.

3 Place the pan in the oven, and roast until the chicken is dark brown, 1½ to 2 hours, turning the bones once or twice.

4 Preheat the water bath to 194°F.

5 Remove the chicken from the oven and cool slightly.

6 Add the roasted chicken parts, garlic, and onion, plus the carrots, thyme, bay leaves, peppercorns, and water to a 2-gallon zip-top bag.

7 Pour a little tap water into the hot roasting pan and scrape up any caramelized bits. Add them to the zip-top bag. Seal the bag using the displacement method.

8 Place the bag in the water bath and cook for 12 hours.

9 Remove the bones using tongs and discard. Strain the stock, discarding the solids. Season with salt and pepper. If not using immediately, chill the stock in an ice bath for 30 minutes, then move to the refrigerator or freezer. The fat is easier to remove from the stock once chilled.

Hollandaise Sauce

YIELD: 4 SERVINGS / ACTIVE TIME: 15 MINUTES / TOTAL TIME: 1 HOUR

Hollandaise sauce is one of the five French "mother sauces" and is notoriously tricky. Because you're dealing with eggs, butter, and heat, it's easy to see the sauce break, curdle, or scramble before your eyes. Sous vide cooking allows you to perfectly dial in the temperature without risking overcooking. You can even keep the sauce warm until the rest of your dish is ready to serve.

¾ cup dry white wine
2 tablespoons champagne or wine vinegar
1 tablespoon minced shallots
1 thyme sprig
4 large egg yolks
1½ sticks (¾ cup) unsalted butter, melted, warm
½ tablespoon freshly squeezed lemon juice
Salt
Freshly ground black pepper

1. Preheat the water bath to 149°F.
2. In a medium saucepan, bring the wine, vinegar, shallots, and thyme to a boil. Reduce the heat and simmer for about 10 minutes.
3. Strain the wine mixture through a fine-mesh strainer into a blender. Add the egg yolks and blend until smooth and frothy, about 30 seconds.
4. Transfer the mixture to a canning jar or large zip-top bag. For the canning jar, close the lid just tight enough so any air will be able to escape from the jar when submerged in water. If you close it too tightly, the trapped air will press against the glass and could crack or break your jar. For the zip-top bag, seal using the displacement method.
5. Place the mixture in the water bath and cook for 45 minutes, shaking the jar or agitating the bag midway through the cooking time.

6 Remove the jar or bag from the water bath.

7 Return the contents of the jar or bag to the blender and add the warm butter and lemon juice.

8 Purée until smooth and frothy, about 30 seconds. Season with salt and pepper and serve. An immersion blender works just as well as a regular blender for this.

Tip: *Béarnaise sauce is an offshoot or "child sauce" of hollandaise, the main difference being the addition of shallot, chervil, peppercorns, and tarragon. This hollandaise recipe adds shallot and thyme, putting it somewhere between its classic form and a béarnaise. For a traditional hollandaise, omit the shallot and thyme, and for a stepped-up béarnaise, add chervil and tarragon to the recipe.*

Mushroom Broth

YIELD: 1 QUART / ACTIVE TIME: 10 MINUTES / TOTAL TIME: 2½ HOURS

Fresh mushrooms are great for a stock, but dried mushrooms are full of rich, concentrated flavors. I like to use a combination of both to create a flavorful broth that is a wonderful base for soups, pan sauces, and risotto.

- **2 teaspoons olive oil**
- **1 cup fresh shiitake (or your favorite) mushrooms**
- **2 scallions, thinly sliced, white and green parts separated**
- **2 garlic cloves**
- **1 (1-inch) piece ginger (optional)**
- **1 tablespoon sesame oil**
- **1 cup dried porcini (or your favorite) mushrooms**
- **3 cups of water**

1 Preheat the water bath to 150°F.

2 In a large skillet, heat the olive oil over medium-high heat until hot. Add the fresh mushrooms and cook, stirring occasionally, until browned and crispy, 4 to 6 minutes.

3 Add the white bottoms of the scallions, garlic, ginger (if using), and sesame oil. Cook, stirring occasionally, for 30 seconds to 1 minute, or until fragrant.

4 Add the fresh mushroom mixture, porcini, greens from the scallions, and water to a zip-top bag. Seal using the displacement method.

5 Place the bag in the water bath and cook for 2 hours.

6 Prepare an ice bath.

7 Remove the bag from the water bath and place it in the ice bath for 15 to 20 minutes.

8 Strain the broth, pressing on the mushrooms to release all their juices, and discard the solids. Store the broth in a sealed container in the refrigerator or freezer.

Tomato Sauce

YIELD: 1 CUP / ACTIVE TIME: 20 MINUTES / TOTAL TIME: 1½ HOURS

Tomato sauce is traditionally cooked down for hours on a stove top or in the oven to allow all the flavors to mingle and marry together into a rich, mellow brew. The benefits of cooking it at a lower temperature sous vide are that you preserve all of the bright, fresh flavors that are lost at higher heats but still retain the deep, hearty notes of a traditional sauce, and in half the time. You can use any tomatoes, but San Marzanos are my favorite. Romas are also a good choice.

2 tablespoons olive oil, divided
½ cup chopped shallots
½ cup chopped onion
2 garlic cloves, sliced
2 pounds ripe tomatoes
3 fresh oregano sprigs
3 fresh thyme sprigs
6 large basil leaves, chopped
⅓ cup chopped parsley leaves
Salt
Freshly ground black pepper

1 Preheat the water bath to 183°F.

2 Heat 1 tablespoon of oil in a large cast iron or stainless steel skillet over medium heat until the oil is slightly shimmering. Add the shallots, onion, and garlic and sauté for 5 to 7 minutes. (This step is optional since we are cooking the sauce at 183°F, which will allow the vegetables to cook, but I prefer the richness this adds to the flavor.)

3 Scrape the onion mixture into a quart-size vacuum-seal bag or zip-top bag. Add the tomatoes, oregano, thyme, basil, and parsley to the bag. Seal the bag, keeping the tomatoes in a single layer for even cooking, using either a vacuum sealer or the displacement method.

4 Place the bag in the water bath to cook for 45 minutes to 1 hour.

5 Remove the bag from the water, open, and let cool for a few minutes.

6 Carefully remove the skins from the tomatoes and discard. Remove the oregano and thyme sprigs and discard.

7 Place the peeled tomatoes and the remaining contents of the bag in a food processor and process to the desired texture. Season with salt and pepper.

Tip: *Make a double batch in separate bags and freeze half for later!*

Sweet and Spicy Barbecue Sauce

YIELD: 2 CUPS / ACTIVE TIME: 30 MINUTES / TOTAL TIME: 30 MINUTES

This classic barbecue sauce is perfect for pulled beef burgers, but also great on ribs, chicken, and meatloaf. I love to play with the balance of sweet and spicy by adding more or less of the chili and molasses based on how I'm feeling that day. Make up a big batch to keep in the refrigerator for all your barbecue cravings.

2 teaspoons olive oil
½ medium onion, finely chopped
2 garlic cloves, minced
1 tablespoon chili powder
½ teaspoon freshly ground black pepper
1 teaspoon salt
1 cup ketchup
¼ cup yellow mustard
¼ cup apple cider vinegar
2 tablespoons Worcestershire sauce
2 tablespoons molasses
¼ cup dark or light brown sugar

1 In a medium saucepan, heat the oil over medium heat until it shimmers. Add the onion and garlic and sauté until fragrant, about 30 seconds.

2 Add the chili powder, pepper, salt, ketchup, mustard, vinegar, Worcestershire sauce, molasses, and sugar. Reduce the heat to low and stir occasionally until the sauce has thickened, about 20 minutes. Do not let the sauce boil.

Classic Hot Wing Sauce

YIELD: 2 CUPS / ACTIVE TIME: 10 MINUTES / TOTAL TIME: 10 MINUTES

If you've had Buffalo wings before, chances are good that you've had a version of this sauce. Frank's hot sauce is said to be the original, but most hot sauces will work fine—it's easy to create your own version by using different brands.

1 (12-ounce) bottle hot pepper sauce, such as Frank's
½ cup (1 stick) butter
½ teaspoon Worcestershire sauce
½ teaspoon Tabasco sauce

1 In a medium saucepan over low heat, add the hot pepper sauce and butter and cook, stirring occasionally, until the butter is melted.

2 Stir in the Worcestershire and Tabasco sauces.

Spice Rubs

Having a collection of great spice rubs on hand is a surefire way not only to elevate the flavors in your dishes but also to provide variety. You can buy premixed blends, but for optimal freshness and customization, I recommend mixing up your own. You can use whole spices and grind them up or use preground if that's not an option.

For each of the rub mixtures:
In a small bowl, mix together all the ingredients. Store in an airtight container in your pantry for up to 1 year.

Pro tip: *Bloom your spices. Toasting many spices amplifies their flavors by releasing their oils. So here's a trick to take your rubs to the next level: Warm a frying pan over medium heat and pour in your spices. Stir or shake them often. Don't let them sit still for more than 10 seconds or they can burn. It takes only 1 to 2 minutes to bloom them. You'll know when the fragrance jumps out at you.*

STEAKHOUSE RUB

YIELD: ABOUT ½ CUP

This is a great standard rub for steak, pork chops, and chicken.

2 tablespoons chili powder
1 tablespoon smoked paprika
1 tablespoon dried thyme
1 tablespoon dried oregano
1 tablespoon freshly ground black pepper
1 tablespoon celery salt
1 tablespoon garlic powder
1 tablespoon onion powder
1 teaspoon salt

SMOKY SPICE RUB

YIELD: ABOUT ½ CUP

I use this rub on just about anything I want to remind me of a barbecue. It's great on ribs, chicken, and heartier fish.

2 tablespoons freshly ground black pepper
2 tablespoons chipotle powder
2 tablespoons smoked paprika
1 tablespoon mustard powder
2 teaspoons onion powder
2 teaspoons garlic powder

SWEET SPICE

YIELD: ABOUT ½ CUP

Slightly exotic, sweet, and spicy, this is a great mix to add some warm, earthy notes to pork, chili, or even mulled wine.

2 tablespoons star anise
2 tablespoons ground ginger
1 tablespoon freshly ground black pepper
1 tablespoon ground cinnamon
2 teaspoons dried orange peel
2 teaspoons nutmeg
1 vanilla bean, split and scraped

NAPA VALLEY RUB

YIELD: ABOUT ½ CUP

Another standard, this rub is wonderful on everything, from meats to vegetables to eggs.

2 tablespoons sun-dried tomato powder
2 tablespoons rosemary
1 tablespoon ground dried lemon peel
1 tablespoon ground fennel seed
2 teaspoons freshly ground black pepper
2 teaspoons chili powder
2 teaspoons garlic powder
2 teaspoons sea salt

continued ›

SWEET AND SPICY RUB

YIELD: ABOUT ⅓ CUP

This one's a lot like the Smoky Spice Rub (page 215) with the added hint of sweetness from the brown sugar. I love it on brisket and ribs.

3 tablespoons freshly ground black pepper
1 tablespoon onion powder
1 tablespoon brown sugar
2 teaspoons mustard powder
2 teaspoons garlic powder
1 teaspoon chili powder
1 teaspoon chipotle powder
1 teaspoon cayenne pepper

MEMPHIS DUST

YIELD: ABOUT ½ CUP

The Memphis Dust Rib Rub is an incredibly popular rub that has been written up all over the place, and for good reason. Here's my take on it:

¼ cup firmly packed dark brown sugar
2 tablespoons smoked paprika
1 tablespoon garlic powder
2 teaspoons freshly ground black pepper
2 teaspoons ground ginger
1 teaspoon onion powder
½ teaspoon salt
¼ teaspoon ground rosemary
¼ teaspoon dried oregano

FISH SEASONING

YIELD: ABOUT ½ CUP

This blend is great on both lighter fish and green vegetables. I have a friend who swears by it and puts it on everything!

1 tablespoon salt
1 tablespoon sugar
1 tablespoon freshly ground black pepper
1 tablespoon dried parsley
1 tablespoon dried basil
1 tablespoon onion powder
1 tablespoon garlic powder
1 tablespoon ground dried lemon peel
1 teaspoon smoked paprika

CHINESE FIVE SPICE

YIELD: ⅓ CUP

Chinese Five Spice powder adds Asian flair to pork belly, chicken, and anything else you want to add some sweet heat to.

1 tablespoon ground cinnamon
1 tablespoon ground cloves
1 tablespoon toasted ground fennel seed
1 tablespoon ground star anise
1 tablespoon toasted ground Szechuan peppercorns or black pepper

continued ›

RAS EL HANOUT

YIELD: ABOUT ½ CUP

A classic Moroccan spice blend, this lends heat and depth to meats, fish, and grains.

2 tablespoons ground cumin
2 tablespoons ground ginger
2 tablespoons salt
1 teaspoon freshly ground black pepper
1 teaspoon cayenne pepper
1 teaspoon ground allspice
1 teaspoon ground cinnamon
1 teaspoon ground coriander seeds

ROTISSERIE CHICKEN RUB

YIELD: ½ CUP

Mimicking your classic rotisserie chicken flavors, this is a great blend for poultry or added to rice.

1 tablespoon salt
1 tablespoon garlic powder
1 tablespoon ground dried lemon peel
1 tablespoon dried parsley
1 tablespoon dried rosemary
1 tablespoon dried thyme
1 tablespoon freshly ground black pepper
1 tablespoon chili powder

LAMB RUB

YIELD: ½ CUP

The mint and lemon cut through the richness of dishes like lamb or duck, but this is just as good with pork as it is with game.

2 tablespoons dried rosemary
1 tablespoon garlic powder
1 tablespoon chili powder
1 tablespoon paprika
2 teaspoons mustard powder
2 teaspoons dried mint
1 teaspoon ground dried lemon peel
1 teaspoon onion powder
1 teaspoon freshly ground black pepper
1 teaspoon salt

ALL-PURPOSE GAME RUB

YIELD: ABOUT ½ CUP

Great for duck, venison, elk, and anything with deep, earthy flavors.

2 tablespoons dark brown sugar
1 tablespoon coarse salt
1 tablespoon garlic powder
1 tablespoon paprika
2 teaspoons freshly ground black pepper
2 teaspoons onion powder
2 teaspoons dried rosemary
1 teaspoon dry mustard
1 teaspoon ground sage
1 teaspoon cayenne pepper

TIME AND TEMPERATURE COOK CHART

	TEMP (°F)	MIN TIME	MAX TIME
Eggs			
Poached	145	45 minutes	2 hours
Hardboiled	165	45 minutes	2 hours
Poultry			
Light Meat	146	1 hour	3 hours
Dark Meat	165	2 hours	6 hours
Pork			
Tender Cuts	140	1.5 hours	4 hours
Tough Cuts	140	12 hours	48 hours
Beef—Steaks			
Rare	128	1 hour	2 hours
Medium-Rare	134	1 hour	2 hours
Medium	144	1 hour	2 hours
Medium-Well	150	1 hour	2 hours
Well	155	1 hour	2 hours
Beef—Tough Cuts			
Medium-Rare	130	8 hours	24 hours
Medium	140	8 hours	24 hours
Medium-Well	150	8 hours	24 hours
Well	160	8 hours	36 hours
Seafood			
Fish	122	40 minutes	1.5 hours
Shellfish	130	15 minutes	45 minutes
Vegetables			
Tender	183	15 minutes	1 hour
Tough	183	1 hour	3 hours

THE DIRTY DOZEN & CLEAN FIFTEEN

A nonprofit and environmental watchdog organization called Environmental Working Group (EWG) looks at data supplied by the US Department of Agriculture (USDA) and the Food and Drug Administration (FDA) about pesticide residues and compiles a list each year of the best and worst pesticide loads found in commercial crops. You can refer to the Dirty Dozen list to know which fruits and vegetables you should always buy organic. The Clean Fifteen list lets you know which produce is considered safe enough when grown conventionally to allow you to skip the organics. This does not mean that the Clean Fifteen produce is pesticide-free, though, so wash these fruits and vegetables thoroughly.

These lists change every year, so make sure you look up the most recent list before you fill your shopping cart. You'll find the most recent lists as well as a guide to pesticides in produce at EWG.org/FoodNews.

2016 DIRTY DOZEN

Apples

Celery

Cherry tomatoes

Cucumbers

Grapes

Nectarines

Peaches

Potatoes

Snap peas

Spinach

Strawberries

Sweet bell peppers

In addition to the Dirty Dozen, the EWG added two foods contaminated with highly toxic organo-phosphate insecticides:

Hot peppers

Kale/Collard greens

2016 CLEAN FIFTEEN

Asparagus

Avocados

Cabbage

Cantaloupe

Cauliflower

Eggplant

Grapefruit

Kiwis

Mangoes

Onions

Papayas

Pineapples

Sweet corn

Sweet peas (frozen)

Sweet potatoes

MEASUREMENT CONVERSIONS

VOLUME EQUIVALENTS (DRY)

US STANDARD	METRIC (APPROXIMATE)
⅛ teaspoon	0.5 mL
¼ teaspoon	1 mL
½ teaspoon	2 mL
¾ teaspoon	4 mL
1 teaspoon	5 mL
1 tablespoon	15 mL
¼ cup	59 mL
⅓ cup	79 mL
½ cup	118 mL
⅔ cup	156 mL
¾ cup	177 mL
1 cup	235 mL
2 cups or 1 pint	475 mL
3 cups	700 mL
4 cups or 1 quart	1 L
½ gallon	2 L
1 gallon	4 L

VOLUME EQUIVALENTS (LIQUID)

US STANDARD	US STANDARD (OUNCES)	METRIC (APPROXIMATE)
2 tablespoons	1 fl. oz.	30 mL
¼ cup	2 fl. oz.	60 mL
½ cup	4 fl. oz.	120 mL
1 cup	8 fl. oz.	240 mL
1½ cups	12 fl. oz.	355 mL
2 cups or 1 pint	16 fl. oz.	475 mL
4 cups or 1 quart	32 fl. oz.	1 L
1 gallon	128 fl. oz.	4 L

OVEN TEMPERATURES

FAHRENHEIT	CELSIUS (APPROXIMATE)
250°F	120°C
300°F	150°C
325°F	165°C
350°F	180°C
375°F	190°C
400°F	200°C
425°F	220°C
450°F	230°C

REFERENCES

Chapter 1

Christensen, Emma. "How to Sear Meat Properly." The Kitchn. October 21, 2013. http://www.thekitchn.com/how-to-sear-meat-47333

Dvorsky, George. "How to Recognize the Plastics That Are Hazardous to Your Health." March 28, 2013. http://io9.gizmodo.com/how-to-recognize-the-plastics-that-are-hazardous-to-you-461587850

Hesser, Amanda. "Under Pressure." *The New York Times.* August 14, 2005. http://www.nytimes.com/2005/08/14/magazine/under-pressure.html

Logsdon, Gary. "Sous Vide Torch Reviews." Amazing Food Made Easy. Accessed September 15, 2016. http://www.amazingfoodmadeeasy.com/info/modernist-equipment/more/best-sous-vide-torch-reviews

Lopez-Alt, J. Kenji. "Cook Your Meat in a Beer Cooler: The World's Best (and Cheapest) Sous Vide Hack." Serious Eats. April 19, 2010. http://www.seriouseats.com/2010/04/cook-your-meat-in-a-beer-cooler-the-worlds-best-sous-vide-hack.html

Chapter 2

Gruber, Peter. "Impact of Bag Position on Temperature Uniformity." Sous Vide Wikia. July 10, 2010. http://sousvide.wikia.com/wiki/Impact_Of_Bag_Position_On_Temperature_Uniformity

Logsdon, Jason. "Sous Vide Safety: Salmonella and Bacteria." Amazing Food Made Easy. Accessed September 15, 2013. http://www.amazingfoodmadeeasy.com/info/sous-vide-safety/more/sous-vide-safety-salmonella-and-bacteria

Molecular Recipes.com. "Preparing Food for Sous Vide Cooking." Accessed September 15, 2016. http://www.molecularrecipes.com/sous-vide-class/preparing-food/

Chapter 3

Goldwyn, Meathead. "The Science of Rubs." AmazingRibs.com. Accessed September 15, 2016. http://amazingribs.com/recipes/rubs_pastes_marinades_and_brines/the_science_of_rubs.html

Logsden, Jason. "How to Sous Vide Shellfish." Amazing Food Made Easy. Accessed September 15, 2016. http://www.amazingfoodmadeeasy.com/sous-vide-times-temperatures/shellfish

ABOUT THE AUTHOR

In 2014, Sarah James was given a sous vide unit and quickly fell in love with the technique. Encouraged by early success, she quickly made it her go-to method for simple, weekday evening meals. She's spent years learning about the science behind it and having fun re-creating her favorite recipes sous vide. Sarah has published hundreds of online tutorials covering everything from cooking and sewing to laser cutters and wearable electronics. She blogs at www.sousvidely.com.

ACKNOWLEDGMENTS

To Peter who got me hooked in the first place.

To Kieran who encouraged me to follow this passion.

To Mom who was willing to sample my experiments.

To Keely and Maggie who helped me get the words on the page.

To everyone at Callisto Media for inviting me on this journey and making my dream a reality.

RECIPE INDEX

INDEX

D

E

F

CPSIA information can be obtained
at www.ICGtesting.com
Printed in the USA
BVOW11s0037101116
467293BV00003B/3/P